Lecture Notes in Biomathematics

Managing Editor: S. Levin

13

Mathematical Models
in Biological Discovery

Edited by D. L. Solomon and C. Walter

Springer-Verlag
Berlin · Heidelberg · New York 1977

Editors

Daniel L. Solomon
Biometrics, Unit, Cornell University
Ithaca, New York 14853/USA

Charles F. Walter
Department of Chemical Engineering
University of Houston
Houston, Texas 77004/USA

QH
323.5
.M365

Library of Congress Cataloging in Publication Data
Main entry under title:

Mathematical models in biological discoveries.

(Lecture notes in biomathematics ; 13)
Based on papers presented at a symposium entitled
"Contributions of mathematical models to biological
discovery" held during the 141st annual meeting of
the American Association for the Advancement of
Science in New York during January, 1975.
1. Biology--Mathematical models--Congresses.
I. Solomon, Daniel L., 1941-   II. Walter,
Charles, 1936-   III. American Association for
the Advancement of Science.   IV. Series.

Library of Congress Cataloging in Publication Data
QH323.5.M365   574'.01'84   77-1396

AMS Subject Classifications (1970): 92-03, 92A05, 92A10, 92A15, 80A30

ISBN 3-540-08134-8   Springer-Verlag Berlin · Heidelberg · New York
ISBN 0-387-08134-8   Springer-Verlag New York · Heidelberg · Berlin

© by Springer-Verlag Berlin · Heidelberg 1977
Printed in Germany.

Printing and binding: Beltz Offsetdruck, Hemsbach/Bergstr.
2145/3140-543210

TABLE OF CONTENTS

PREFACE

When I was asked to help organize an American Association for the
ancement of Science symposium about how mathematical models have con-
buted to biology, I agreed immediately. The subject is of immense
ortance and wide-spread interest. However, too often it is discussed
biologically sterile environments by "mutual admiration society"
ups of "theoreticians", many of whom have never seen, and most of
m have never done, an original scientific experiment with the biolog-
l materials they attempt to describe in abstract (and often prejudiced)
ms. The opportunity to address the topic during an annual meeting
the AAAS was irresistable. In order to try to maintain the integrity
the original intent of the symposium, it was entitled, "Contributions
Mathematical Models to Biological Discovery".

This symposium was organized by Daniel Solomon and myself, held
ing the 141st annual meeting of the AAAS in New York during January,
5, sponsored by sections G and N (Biological and Medical Sciences)
the AAAS and the North American Regions of the Biometric Society,
supported by grant BMS 75-02803 from the National Science Foundation.

What follows in this volume are papers by nine of the participants
not only felt that they had something to say in a symposium entitled,
ntributions of Mathematical Models to Biological Discovery", but who
o were willing to record their ideas in more detail here.

The first paper introduces the role of mathematical models in biol-
via historian Provine's elegant summary of the contributions popula-
n genetics models have made to our understanding of evolution. The
t six manuscripts describe specific models and some of their contri-
ions to biological discovery. These six papers are arranged according
the approximate biological heirarchy of the models they deal with:
cellular (enzymes), cellular (mitotic clock), incipient multicellular

(differentiation), multicellular (physiological/biochemical), single

organism (feedback loops associate with backpacking), and multi-orga

(ecosystems). In the penultimate paper, Professor van der Vaart out

his impressions of some biology/mathematics interactions and his ass

ment of the degree to which they have been fruitful. The last paper

contains some speculation about what type of mathematics is most sui

for constructing models in biology.

The volume opens with Professor Provine's treatise about the con-
ributions mathematical models of population genetics have made to our
derstanding of evolution. Of particular interest here are some new
sights into the relationship between Sewall Wright and Theodosius
ozhansky, insights gained during Provine's recent meetings with Wright.

ROLE OF MATHEMATICAL POPULATION GENETICISTS IN THE EVOLUTIONARY SYNTHESIS OF THE 1930'S AND 40'S[1]

by William B. Provine, Department of History, Cornell University

The role played by exact mathematics in the development of scie[nce] is often controversial. When Isaac Newton published his Mathematica[l] Principles of Natural Philosophy, Robert Hooke immediately responded that Newton had merely explored the mathematical consequences of Hoo[ke's] own idea of the inverse square law of gravitational attraction. To [this] accusation, Newton angrily replied:

> Mathematicians, who find out, settle, and do all the business, [must] content themselves with being nothing but dry calculators and drudges; and another, who does nothing but pretend and grasp at all things, must carry away all the invention.[2]

Newton won the argument. The Principia exerted an enormous influenc[e] upon science because Newton's mathematical analysis enabled him to de[scribe] more accurately than most people had dreamed possible, the motions o[f] the planets, the comets, the moon, and the seas. Having achieved th[is] stunning success, Newton, in the preface to the first edition of the Principia, expressed the hope: "I wish we could derive the rest of [the] phenomena of Nature by the same kind of [mathematical] reasoning fro[m] mechanical principles." Since the Principia, mathematical scientist[s] have indeed been motivated by hopes which often closely resemble tha[t] expressed by Newton.

In many fields, however, the triumphs of the mathematical scien[ces] have been less clear-cut than in the case of Newton. The role of ma[th]ematics in the biological and social sciences has been especially co[n]troversial, particularly in areas like evolutionary biology or econo[my] where "mathematical models" have many proponents and detractors.

The rapid development of evolutionary thought in the two decade[s] between 1930 and 1950 has been termed by Julian Huxley, Ernst Mayr, Theodosius Dobzhansky, and others as the "modern synthesis" or more often the "evolutionary synthesis." The synthesis encompassed genet[ics] systematics, paleontology, and cytology. This paper addresses the s[pe]cific question: What role did mathematical population geneticists a[nd] the models they created play in the evolutionary synthesis?

By the early 1950's the answer to this question appeared obvious to most evolutionists. Dobzhansky, whose very influential book Genet[ics] and the Origin of Species had been through three editions (1937, 194[1,]

951), stated in his introductory address to the 1955 Cold Spring Harbor
ymposium on population genetics that

> The foundations of population genetics were laid chiefly by
> mathematical deduction from basic premises contained in the works
> of Mendel and Morgan and their followers. Haldane, Wright, and
> Fisher are the pioneers of population genetics whose main research
> equipment was paper and ink rather than microscopes, experimental
> fields, _Drosophila_ bottles, or mouse cages. This is theoretical
> biology at its best, and it has provided a guiding light for rig-
> orous quantitative experiment and observation.[3]

. M. Sheppard, an Oxford evolutionist who was also present at that
eeting, had a year earlier published an essay in the important 1954
olume _Evolution as a Process_ (edited by Julian Huxley, A. C. Hardy, and
. B. Ford) in which he argued

> The great advances in understanding the process of evolution,
> made during the last thirty years, have been a direct result of the
> mathematical approach to the problem adopted by R. A. Fisher,
> J. B. S. Haldane, Sewall Wright, and others.... The hypotheses
> derived by mathematicians have given a great impetus to experimental
> work on the genetics of populations.[4]

The viewpoint offered by Dobzhansky and Sheppard dominates the liter-
.ture treating the relation of genetics to evolution in the early 1950's.
.lmost every monograph or textbook cited R. A. Fisher, J. B. S. Haldane,
.nd Sewall Wright as the co-founders of modern evolutionary theory.

Only a few evolutionists ventured to publicly question the dominant
riew. At the Symposium of the Society of Experimental Biology at Oxford
.n 1952, C. H. Waddington suggested that the "undoubtedly immense import-
.nce and prestige" of the mathematical theory of evolution might not
leserve such high distinction. The mathematical theories of Fisher,
laldane, and Wright

> did not achieve either of the two results which one normally expects
> from a mathematical theory. It has not, in the first place, led to
> any noteworthy quantitative statements about evolution. The form-
> ulae involve parameters of selective advantage, effective population
> size, migration and mutation rates, etc., most of which are still
> too inaccurately known to enable quantitative predictions to be
> made or verified. But even when this is not possible, a mathemat-
> ical treatment may reveal new types of relation and of process,
> and thus provide a more flexible theory, capable of explaining
> phenomena which were previously obscure. It is doubtful how far

the mathematical theory of evolution can be said to have done th
Very few qualitatively new ideas have emerged from it.[5]
Haldane, who was the only one of the three present at this meeting,
responded by citing some examples in which he suspected that mathemati
models had influenced evolutionary thinking.[6] Haldane's entire defen
comprised only two paragraphs; Fisher and Wright did not take up Wadd
ton's challenge. Waddington republished his 1952 remarks in his 1957
book, The Strategy of the Genes, but no detailed rebuttal appeared.

Ernst Mayr renewed Waddington's challenge and placed it in a his
torical framework in his introductory address at the 1959 Cold Spring
Harbor Symposium on "Genetics and Twentieth Century Darwinism" cele-
brating the Darwin centennial. Mayr divided the history of genetics
relation to evolution into three periods: the Mendelian period (1900-
1920), dominated by a mutation theory of evolution; classical populat:
genetics (1920-late 1930's), characterized by a resurgence of belief
gradual Darwinian natural selection but dominated by the simplistic
view that "evolutionary change was essentially ... an input or output
of genes, as the adding of certain beans to a beanbag and the with-
drawing of others" ("beanbag genetics"); and the period (late 1930's
to 1959) of "newer population genetics ... characterized by an increas
emphasis on the interaction of genes." The work of Fisher, Haldane,
and Wright, Mayr asserted, belonged to and dominated the period of "be
bag genetics." Mayr cited Waddington, and then renewed the challenge
by asking, "What, precisely, has been the contribution of this mathematical school to the evolutionary theory, if I may be permitted to
ask such a provocative question?" He answered his own question: "Pei
haps the main service of the mathematical theory was that in a subtle
way it changed the mode of thinking about genetic factors and genetic
events in evolution without necessarily making any startlingly novel
contributions." This assessment clearly differed from that provided
by Dobzhansky and Sheppard. Added Mayr, "I should perhaps leave it to
Fisher, Wright, and Haldane to point out themselves what they consider
their major contributions."[7]

Mayr chose a timely and prestigious forum for this challenge,
making it difficult to ignore. Wright, who was present at the Symposi
published the next year a review of the Symposium volume.[8] Almost the
entire review was devoted to a refutation and answer to Mayr's introductory address. Wright countered Mayr's periodization with many examples, and argued convincingly that his own theory of evolution, whic
from the beginning emphasized interaction effects of genes, could not
simply be labelled as "beanbag genetics." But in direct response to

Mayr's query about the actual role played by the mathematical theory in the development of the evolutionary synthesis, Wright responded with only two paragraphs in the eight page review.

Fisher and Haldane did not attend the 1959 Symposium. Fisher never answered Mayr's challenge, but Haldane eventually published his response in 1964 as a popular essay entitled "A Defense of Beanbag Genetics." Haldane considered himself the last bulwark of defense against Mayr's challenge:

Fisher is dead, but when alive preferred attack to defense. Wright is one of the gentlest men I have ever met, and if he defends himself, will not counterattack. This leaves me to hold the fort.[9]

All its wit and examples notwithstanding, this essay is suggestive but provides no adequate analysis of the role played by mathematical theory in the evolutionary synthesis of the 1930's and 40's.

Mayr and Waddington considered the responses of Wright and Haldane to their challenge to be inadequate, and both renewed the challenge at the 1966 Symposium of the Wistar Institute on "Mathematical Challenges to the Neo-Darwinian Interpretation of Evolution."[10] Wright again replied briefly. I have made in this paper a preliminary evaluation of the challenge of Mayr and Waddington, examining in some historical detail the role played by mathematical population genetics in the evolutionary synthesis.

Background to the Mathematical Models of Population Genetics[11]

Charles Darwin produced no quantitative models for his theory of evolution by natural selection, nor did his nineteenth century followers like Alfred Russel Wallace, E. B. Poulton, August Weismann, G. J. Romanes, and many others. Darwin's repeated claim that evolution proceeded by natural selection operating upon the very small variations to be found in every population of organisms, was far from obvious to many experimental biologists and naturalists. Like T. H. Huxley and Francis Galton, these biologists believed that natural selection could not act effectively upon such small variations. By the end of the nineteenth century a growing number of mostly experimental biologists rejected Darwin's theory of gradual evolution, and espoused one of several theories of discontinuous evolution. The division between these two schools of thought was clear and already widening before 1900.

The biometricians Karl Pearson and W. F. R. Weldon did develop quantitative techniques for measuring changes in variability over time in populations, and Weldon even conducted one unsatisfactory study of

natural selection in a population of crabs subjected to muddy water,
their work provided no way of ascertaining whether such changes were
hereditary.

When Mendelism was rediscovered in 1900, adherents of discontinue
evolution, like William Bateson and Hugo de Vries, immediately pointe
out the nice way Mendelian heredity complemented their theory of evolution. Naturalists who believed in a gradual process of evolution,
whether neo-Darwinians or neo-Lamarckians, naturally reacted against
Mendelism and the new science of genetics. For example, older natural
ists like Alfred Russel Wallace, E. B. Poulton, and E. Ray Lankester
had found no de Vriesian mutations in natural populations, and public]
belittled the significance of Mendelism for evolution. Even younger
naturalists like the American Francis B. Sumner reacted similarly. Th
experimental geneticists, as they developed their science during its
first twenty years, were largely isolated from the taxonomists who
studied living populations in nature. The biometricians, who had deve
loped some very useful mathematical methods for studying populations,
fought furiously with the Mendelians. Naturalists, who also disliked
Mendelism, could not understand the significance of the biometricians'
endless correlation tables. Paleontologists adhered mostly to neo-
Lamarckian or discontinuous theories of evolution, sometimes a combination of the two, and had little use for the new science of genetics.
Thus the divisions between the disciplines which would later be synthesized into the modern theory of evolution were quite pronounced dur
the first twenty to thirty years of the century.

Two distinguished American evolutionists, David Starr Jordan and
Vernon L. Kellogg, presented the dilemma cogently in their 1907 book,
Evolution and Animal Life. First they quoted from an essay entitled
"The Unknown Factors of Evolution" by Henry Fairfield Osborn:

> The general conclusion we reach from a survey of the whole field
> is that for Buffon's and Lamarck's factors we have no theory of
> heredity, while the original Darwinian factor, or Neo-Darwinism,
> offers an inadequate explanation of evolution. If acquired var-
> iations are transmitted, there must be, therefore, some unknown
> factor in evolution.

And they added:

> Our present plight seems to be exactly this: we cannot explain
> to any general satisfaction species-forming and evolution without
> the help of some Lamarckian or Eimerian factor; and on the other
> hand, we cannot assume the actuality of any such factor in the
> light of our present knowledge of heredity. The discovery of the

"unknown factors of evolution" should be the chief goal of all present-day biologic investigation.[12] Their major argument against Darwin's theory was, of course, that natural selection must be ineffective acting upon the small variations found in natural populations. Many naturalists simply adhered to the option of neo-Lamarckism and/or directed evolution and thus denied the existence of unknown factors of evolution.

Between 1900 and 1920, however, experimental geneticists furnished much of the groundwork for a possible reconciliation of genetics and gradual Darwinian evolution. They showed that Mendelian heredity preserved variability in populations, that continuous variability could be interpreted as Mendelian by hypothesizing the existence of multifactorial inheritance, that sexual recombination could produce a great amount of heritable variability, and that artificial selection could change a population beyond the limits of variability found initially. Geneticists like Erwin Baur, William Castle, Charles Benedict Davenport, Edward Murray East, Herbert Spencer Jennings, Thomas Hunt Morgan, and many of their students came to realize by 1918 that in a general way Mendelism and Darwinism were complementary. But these geneticists could not easily dispel the argument that small Mendelian variations furnished too small a handle for natural selection. Some of them, like Baur and Morgan, often deemphasized the role of natural selection and emphasized the magnitude of the mutational leaps involved in the process of evolution in nature. Naturalists, with a few exceptions like Julian Huxley, were unimpressed with the work of geneticists. They were quite certain that intense selection like that created by Castle in his experiments on rats, or by A. H. Sturtevant on <u>Drosophila</u>, was rare in the natural populations they studied. By about 1918, however, the stage was prepared for a quantitative theoretical analysis of the consequences of Mendelian heredity in relation to selection and other evolutionary processes.

The exploration of the quantitative consequences of Mendelism had already begun before 1918. The Hardy-Weinberg equilibrium principle was well known, and some rather crude investigations into the mathematical consequences of inbreeding had been carried out. Mathematician H. T. J. Norton had published his important table showing the effects of selection of several intensities acting upon a Mendelian factor in a large random breeding population. Beginning here, between 1918 and 1932, four individuals produced mathematical models for genetic change in populations of organisms. They are Sergei Chetverikov, R. A. Fisher, J. B. S. Haldane, and Sewall Wright.

Only a very brief description of the models these men developed can be given here. Starting from an analysis of the stable distributi of gene frequencies expected in a large random breeding population wit separated generations, they analysed what might happen to a population subject to a wide variety of factors which could conceivably change th frequency of genes predicted by the simple Hardy-Weinberg equilibrium principle. These factors included selection, dominance, mutation, epistasis, population structure, breeding structure, linkage, balanced polymorphisms, random processes, and group selection. In all cases simplifying assumptions were necessary in order to make the models man ageable.

Fisher, Haldane, and Wright emphasized and developed different aspects of the evolutionary process in their models. Initially, Fishe was most interested in the effects of low selection pressures on in- dividual genes in large populations, Haldane with considerably higher selection rates on individual genes in large populations, and Wright with intermediate selection rates on gene interaction systems, created partly by random genetic drift in small, partially isolated population After the early 1930's, when consistent differences in the selective values postulated by the three men disappeared, other differences emer All three might easily calculate the same numerical answer for the re- sult of a given set of assumptions. But they emphasized different aspects of this evolutionary process, and their initial assumptions generally differed. They therefore predicted different outcomes.

For example, their views on dominance differed markedly. Fisher thought dominance had evolved by low selection pressures operating upo modifiers of heterozygotes that were initially intermediate in pheno- typic expression between the homozygotes. Over a very long period of time, the heterozygote would be selected to look like the homozygous wild type. Because he thought heterozygotes were represented by low frequencies in a natural population, Fisher postulated selection pres- sures of the order of mutation pressures, which were very low. Wright disagreed with this theory, arguing that such low selection pressures would be lost amid higher selection pressures operating upon the same modifiers. Wright reasoned that any modifier of the heterozygote woul probably also affect some characters in the abundant homozygous in- dividuals, thus generating higher selection pressures upon Fisher's hypothetical modifiers than resulted from their effects upon the more rare heterozygotes. From his own experiments on guinea pigs, Wright was fully aware that selection could modify dominance, even reverse it, but he believed that similar conditions would rarely occur in natur

nstead he suggested a physiological interpretation of gene action in
ominance, concluding that a new mutant appearing in a pristine heter-
zygote was likely to be fully recessive to the wild type allele from
he beginning, rather than intermediate as Fisher argued. Haldane
haracteristically proposed a model in which selection acted deter-
inistically upon single alleles, with far greater pressures than Fisher
ad hypothesized.

The differences between the models proposed by Fisher, Haldane,
nd Wright created tensions with several different effects. Most im-
ortantly, the tensions were scientifically fruitful by stimulating many
eneticists to study natural populations quantitatively, and by causing
ach of the three thinkers to modify his own views over time. But the
ensions also had negative effects. Fisher was unable to graciously
ccept Wright's criticisms of his views and communications between them,
igorous in 1929-30, soon broke down entirely. The failure in private
ommunications led to sometimes bitter disagreements in the journals.
nother effect of these tensions relating directly to the evolutionary
ynthesis was that some naturalists, who generally had little preparation
n mathematics and who were already suspicious of the simplifying as-
umptions made by the mathematical workers, found that the arguments
etween Fisher, Haldane, and Wright reinforced their suspicion that the
athematical models were useless for understanding the complexities of
volution in natural populations.

The mathematical models were more than abstract mathematical re-
ations, but they were less than descriptions of the dynamics of natural
opulations. Fisher, Haldane, and Wright each began by identifying the
elevant variables, including selection, dominance, etc., mentioned above.
any of these variables had been studied, sometimes carefully, by natural-
sts or geneticists. Fisher, Haldane, and Wright then framed hypotheses
bout the relations of these biological variables, and introduced simpli-
ications so that the relations could be treated by manageable mathe-
atics. The mathematical relations thus produced could be solved or
xtrapolated to <u>possible</u> simplified descriptions of the dynamics of
atural populations. These possible simplified descriptions often had
estable consequences in natural populations; the magnitude of the bio-
ogical variables, and the hypothesized relations between them, were
lso subject to various kinds of experimental investigation.

Methods

I first collected the books and files of all the major papers of

Fisher, Haldane, and Wright. I read through these chronologically, taking careful notes on related literature they cited. Then I read through about twenty-five of the major books on evolution and genetics published between 1932 and 1959. These included such works as Th. Dobzhansky's <u>Genetics and the Origin of Species</u>, Julian Huxley's <u>Evolution: the Modern Synthesis</u> and (edited) <u>The New Systematics</u>, Ernst Mayr's <u>Systematics and the Origin of Species</u>, G. G. Simpson's <u>Tempo and Mode in Evolution</u>, Bernhard Rensch's <u>Neuere Probleme der Abstammungslehre</u>,[13] and the Cold Spring Harbor Symposia for 1955 on population genetics and for 1959 on genetics and twentieth century Darwinism. Here again I took notes on related literature in the journals. And finally I scoured this journal literature, finding of course more pertinent references. The analyses of the influence of mathematical population genetics written by Haldane and Wright, although brief, were very suggestive and helpful. This procedure was reasonably effective for elucidating the major lines of influence of the mathematical models, as found in published sources.

The published literature, however, can be misleading and does not tell the whole story. For example, simple citation of the writings of Chetverikov, Fisher, Haldane, and Wright is insufficient reason to assume any significant influence of their quantitative work upon the person citing it. I have therefore utilized several other sources to supplement the published literature. The most important source was the unpublished correspondence of persons involved in the evolutionary synthesis, or affected by it. The most helpful single collection was the striking correspondence between Sewall Wright and Theodosius Dobzhansky in the years 1937-1950, which Wright kindly made available to me. Also helpful were the formal and informal contributions of many participants in a conference on the evolutionary synthesis sponsored by Ernst Mayr and the American Academy of Arts and Sciences in 1974. Participants who provided insights about the mathematical work include Dobzhansky, C. D. Darlington, E. B. Ford, I. Michael Lerner, R. C. Lewontin, and Mayr. Julian Huxley, Bernhard Rensch, G. G. Simpson, and M. J. D. White, although they did not attend the conference, made written contributions. Extensive personal interviews with Sewall Wright provided invaluable insights. My research is still far from complete, and the results reported below are tentative.

Brief Statement of Results

My research to date indicates a clear conclusion to the question

sed at the beginning of this paper. The mathematical models of pop-
ation genetics had a significant influence upon evolutionary thinking
 at least four ways. First, these models demonstrated to most bio-
gists in the 1930's and 40's that Mendelism and natural selection,
us processes known or reasonably supposed to exist in natural popula-
ons, were sufficient to account for micro-evolution at the population
vel. Second, the models indicated that some paths taken by evolution-
y biologists were not fruitful. Third, the models elucidated, comple-
nted, and lent greater significance to the results of field researches
ready completed or in progress. These researches included the work
 systematists and paleontologists, as well as geneticists. Fourth,
e models stimulated and provided an intellectual framework for later
eld research. And finally I conclude that the influence of the models
 greater than many biologists think, because their impact was some-
mes invisible to those influenced. I will amplify each of these con-
usions in following sections.

Sufficiency of Mendelism and Selection

From 1859 on there had always been Darwinians who believed that
e primary mechanism of evolution was natural selection operating upon
all observable variations, to some extent hereditary. A varying, but
lways far greater number of biologists rejected Darwin's conception
f the mechanism of evolution because (among other reasons) they did
ot believe that natural selection could operate effectively upon the
all variations emphasized by Darwin. Between 1900 and 1920 geneticists
ad demonstrated that "continuous variability" could have a hereditary
endelian basis, and some geneticists, like East and Castle, were strict
arwinians. Yet, as other geneticists and biologists continually pointed
ut, Darwinism still rested upon the apparently shaky theoretical claim
hat selection in nature really was effective when acting upon very small
enetic differences between individuals.

Probably the greatest general contribution of the mathematical
odels of Chetverikov, Fisher, Haldane, and Wright was to put Darwin's
dea of gradual evolution by natural selection on a firm theoretical
asis. Here was a solution to the dilemma posed by Jordan and Kellogg
n 1907. At least in a general way, the models showed that Mendelian
eredity, when combined with plausible assumptions about population
tructure, breeding structure, and natural selection, provided a con-
istent, intellectually appealing solution to the problem of the mech-
nism of evolution. Acquired characters were unneeded. There were no

important "unknown factors of evolution," even if some of the factors
known to play a role, such as ecology or development, could not be fu
quantified. Perhaps more importantly, the theoreticians helped resol
the hot debate between adherents of discontinuous evolution and the a
herents of continuous evolution. During the 1930's there emerged for
the first time a general picture of the evolutionary process which a
wide spectrum of biologists found intellectually satisfying.

The evidence for my assertion here is abundant, and comes from
many different sources. Chetverikov, Fisher, Haldane, and Wright all
explicitly say that their work in the late 1920's and early 1930's wa
designed to construct a coherent and scientifically plausible theory
of evolution by synthesizing Mendelian genetics with possible forces
which could change gene frequencies, primarily selection. The openin
sentence of J. B. S. Haldane's first paper on "A Mathematical Theory
Natural and Artificial Selection" (1924) reveals the intent shared by
all four workers:

> A satisfactory theory of natural selection must be quantitativ
> In order to establish the view that natural selection is capable
> of accounting for the known facts of evolution we must show not
> only that it can cause a species to change, but that it can caus
> it to change at a rate which will account for present and past
> transmutations.[14]

The key to the mathematical demonstration of the effectiveness of sel
tion was of course tied to the regularities of Mendelian heredity, so
all four synthesized Mendelism with selection theory. All four had
worked in a general atmosphere in which to varying extents Mendelism
was considered to be antagonistic to Darwinism, and thus all shared
also Fisher's motivation for writing his Genetical Theory of Natural
Selection in 1930. He said that he wrote the book to demonstrate how
little basis there was for the opinion "that the discovery of Mendel'
laws of inheritance was unfavorable, or even fatal, to the theory of
natural selection."[15] So the creators of the models all wanted to
demonstrate at least the possibility of evolution along the lines sug
gested by Charles Darwin.

Both Mayr and Waddington, who believe that many geneticists have
claimed too much for theoretical population genetics, admit the mathematical models had some influence upon evolutionary thought. Thus
Waddington wrote that "the outcome of the mathematical theory was, in
the main, to inspire confidence in the efficiency of the process of
natural selection and in the justice of applying this type of argumen
also to the realm of continuous variation."[16] And Mayr said:

It seems to me that the main importance of the mathematical theory was that it gave mathematical rigor to qualitative statements long previously made. It was important to realize and to demonstrate mathematically how slight a selective advantage could lead to the spread of a gene in a population.[17]

I wish to emphasize the importance of the contribution pointed out Waddington and Mayr. Darwin's idea of evolution through the natural lection of small differences found in natural populations was only one ong many plausible hypotheses about the mechanism of evolution. Neo-narckism, in many variations, was popular among systematists through e nineteen-teens and twenties, and into the thirties; many experimental ologists found variations of de Vries' mutation theory appealing. And rwin's idea was, of course, vulnerable to the obvious and repeated jection that the small selection pressures hypothesized were just too effective to be the primary directive agent in evolutionary change. is objection to Darwin's theory was not prompted by the lack of exper-ental evidence demonstrating that small selection pressures were ef-ctive in causing natural populations to change, but by the difficulty olutionists had in believing that small selection pressures could work pidly enough to cause the changes found in fossil records.

This conceptual difficulty with Darwin's small selection pressures s removed by the new mathematical models of selection. Small selection essures did not _appear_ to be capable of changing populations rapidly ough to fit the geological record, but in fact they could. The problem s in the minds of evolutionists, not in the small selection pressures.

I cannot help being reminded here of Darwin's reaction in the _Origin Species_ to the objection that natural selection could not account for e evolution of organs of extreme perfection, such as the eye. Darwin, a passage often quoted by opponents of natural selection, seemed to ree with the objection:

> To suppose that the eye, with all its inimitable contrivances for adjusting the focus to different distances, for admitting different amounts of light, and for the correction of spherical and chromatic aberration, could have been formed by natural se-lection, seems, I freely confess, absurd in the highest possible degree.

t Darwin, of course, added that the problem was in our heads and not the mechanism of natural selection:

> Yet reason tells me, that if numerous gradations from a perfect and complex eye to one very imperfect and simple, each grade being useful to its possessor, can be shown to exist; if further,

the eye does vary ever so slightly, and the variations be inher
which is certainly the case; and if any variation or modificati
in the organ be ever useful to an animal under changing conditi
of life, then the difficulty of believing that a perfect and co
eye could be formed by natural selection, though insuperable by
imagination, can hardly be considered real.[18]

The difficulty with the effectiveness of small selection pressures w
a similar kind of conceptual problem.

Mayr is right that the new mathematical models of evolution "ga
mathematical rigor to qualitative statements long previously made" b
Darwin, Wallace, Weismann, and others. But the mathematical models
played an additional crucial role. Darwin and all Darwinian natural
after him had been unable to mould a consensus among biologists concerning the mechanism of evolution. Despite their discoveries showi
that some continuous variability had a Mendelian basis, and that mos
genes probably had pleiotropic effects, geneticists in the years 190
1925 had not dispelled the widespread belief that natural selection
acting upon very small variations was an ineffective mechanism of ev
tionary change. In combination with the discoveries in genetics, ne
mathematical models proved to be a significant factor enabling biolo
to discriminate between competing hypotheses about the mechanism of
evolution. The mathematical models did not logically falsify non-Da
winian mechanisms of evolution, nor did they prove that gradual natu
selection was the primary mechanism of evolution. The models did, h
ever, clearly and instantly remove the longstanding objection that s
selection pressures must necessarily be ineffective in evolution.

The new theory of natural selection proved irresistably appeali
because it was consistent with genetics, systematics, cytology, and
paleontology. As a bonus, it harmonized with the views expressed by
Charles Darwin, whose work was always prestigious, even when under at
The mathematical demonstration that small selection pressures _could_
plain evolution was very quickly transmuted into the non-mathematica
common sense belief, which anyone could understand, that in fact suc
small selection pressures _did_ explain evolution in nature.

Important evolutionists like Francis B. Sumner, Bernhard Rensch
and Ernst Mayr were all neo-Lamarckians in the mid-1920's. By the m
1930's all three had become neo-Darwinians, who emphasized the impor-
ance of natural selection acting upon small Mendelian differences and
who attributed no role to the inheritance of acquired characters. Re
has commented on this change:

In the same measure as the experiments of geneticists showed tha

nearly all genes have pleiotropic effects and that selection can become effective in the course of some thousands of generations when
the advantage of a variety is only ½%, I gave up all Lamarckian ex-
planations.[19]

r similar reasons many evolutionists gave up a variety of beliefs and
came neo-Darwinians. By 1940 any evolutionist not a neo-Darwinian
s clearly out of step with the times. The same assertion could not
ve been made in 1925, or even 1930.

The influence of the mathematical models can be traced specifically
many cases. R. A. Fisher's theoretical formulations permeate E. B.
rd's evolutionary writings from the moment they first met. The first
ition of Ford's widely read little book Mendelism and Evolution[20]
932) shows the unmistakable influence of Fisher's work. All of the
ucial discussions of blending inheritance, dominance, selection,
lymorphism, and linkage are explicitly based upon Fisher's analyses
these problems. Yet Ford's treatment of evolutionary problems in
is book was non-mathematical; he simply used the results of Fisher's
thematical work.

Another popular book on the new evolutionary views, Julian Huxley's
olution: The Modern Synthesis, first published in 1942, exhibits the
me sort of influence from Fisher, Haldane, and Wright. Almost every
scussion of the various facets of the evolutionary process is placed
.to the theoretical framework provided by one or the other of these
ree. Each of the three is referred to more often than anyone else
. the book. T. H. Morgan, to whom the book is dedicated, is referred
9 times compared with 35 for Fisher, 44 for Haldane, and 37 for
·ight. In Huxley's introduction to The New Systematics, which he edited
. 1939-40, he had stated that "men like Fisher, Haldane, and Wright
.ve recently done great service by pointing out that selection may take
.ny different forms and achieve very different results according to
.e conditions under which it operates.... The use of mathematical
·thods has given a firm deductive basis for selection theory."[21] The
;reat service" rendered by Fisher, Haldane, and Wright is obvious in
.e 1942 book.

Perhaps the most pervasive influence of the mathematical models of
·pulation genetics can be found through the work of Theodosius Dobzhan-
·y, who was first influenced by Chetverikov, and later by Fisher, Hald-
.e, and Wright before publishing his important book Genetics and the
·igin of Species in 1937. Chetverikov, in the mid-1920's, had emphas-
.ed the possible great effects of even small selection pressures, and
.bzhansky became aware of this view from his early associations with
.etverikov. Fisher, Haldane, and especially Wright had a more explicit

impact upon Dobzhansky's evolutionary thinking, judging from the 193
book. His chapters on "Variation in Natural Populations" and "Selec
are cast within the framework of the mathematical models. The chapt
on selection ends with a long quote in which Sewall Wright summarize
his general view of the evolutionary process.

If the mathematical models of Chetverikov, Fisher, Haldane, and
Wright had influenced only Dobzhansky, the effect would have been si
icant. But the effect was much more pervasive because of the enormo
influence exerted by Dobzhansky upon others. In the twentieth centu
the most influential book on evolution is Dobzhansky's Genetics and
Origin of Species. Nearly all evolutionary biologists read it, and
agree it exerted a great effect upon their views of evolution. Prob
the greatest general effect which the mathematical models had was th
the work of Dobzhansky.

Dobzhansky presented only a very small amount of the actual mat
ematical analysis, and this drawn primarily from Wright. But his vi
on the quantitative process of evolution were, as he says explicitly
drawn from the mathematical models. The call to experimentation fou
in the book, which sent so many biologists into the field, urged re-
searchers to investigate problems elucidated by the mathematical mod
Yet one could read this book with a very elementary understanding of
mathematics. Dobzhansky had translated the evolutionary essentials
the mathematical models into a book which biologists could and did r

J. B. S. Haldane understood very well this translation of mathematical models into biological common sense. When Waddington questi
the influence of mathematical models in evolutionary work at the Symposium for the Society of Experimental Biology in 1952, Haldane (as
put it) "defended his rear" by pointing out that many of the more important ideas in the models "have emerged so completely that their o
is forgotten." He cited the case of balanced polymorphisms, adding
"such ideas as these pass rapidly from being mathematical theorems t
being common sense."[22]

In the second edition of Genetics and the Origin of Species (19
Dobzhansky introduced his discussion of Wright's views with the state
ment, "his mathematical arguments are far too abstruse to be presente
here, but his conclusions are simple enough."[23] Dobzhansky could ha
said the same of his treatment of the views of Fisher and Haldane.
tics and the Origin of Species owed part of its marked success to the
skillful transformation of mathematical models into evolutionary idea
biologists could easily understand.

Several evolutionists have argued that they actually read Fisher

ldane, and Wright very late, or not at all, and were therefore per-
nally little influenced by them. Mayr, for example, has said that
 did not read the mathematical evolutionists until after writing his
fluential book Systematics and the Origin of Species (1942), and later
und them not very helpful. But Mayr emphasizes in the 1942 book the
bt he owed to Dobzhansky and N. W. Timofeef-Ressovsky for an under-
anding of genetics in relation to evolution, and both of them have
plicitly acknowledged in detail their debts to the mathematical evolu-
onists. Perhaps what happened in the case of Mayr, and perhaps of
hers, was that they had already absorbed most of the major ideas of
e theoretical population geneticists before reading them.

Not all evolutionary biologists and geneticists found the mathemat-
al models helpful or useful. Sir E. B. Poulton, who was a staunch
liever in Darwinian evolution, and who knew about the work of Fisher
d Ford, simply saw their work as a modern restatement of what Charles
rwin and Henry Walter Bates had said before 1861. In the lead article
 a volume entitled Evolution, edited by Gavin de Beer and published in
38,[24] Poulton never mentioned the word genetics even though the title
 his piece was "Insect Adaptation as Evidence of Evolution by Natural
election." Some geneticists who read Fisher were unaffected, such as
 C. Punnett in England or Raymond Pearl in the United States. But
ese geneticists were known by their colleagues to be behind the times,
d the influence of the mathematical models on the generally accepted
ew of evolution was deeply felt in the ways I have just described.

Elimination of Competing Theories

My second conclusion is that the mathematical evolutionists demon-
trated that some paths taken by evolutionary biologists were unlikely
 be fruitful. Many of the followers of Hugo de Vries, including some
endelians like Raymond Pearl, believed that mutation pressure was the
st important factor in evolutionary change. The mathematical models
learly delineated the relationships between mutation rates, selection
ressure, and changes of gene frequencies in Mendelian populations. Most
volutionists believed that selection coefficients in nature were several
rders of magnitude larger than mutation rates; upon this assumption,
ne mathematical models indicated that under most conditions likely to
 found in natural populations, selection was a vastly more powerful
gent of evolutionary change than mutation. Additionally, Fisher in
930 made a strong mathematical argument that as the magnitude of an
ndirected mutational change increased, the probability of improving

adaptation rapidly decreased. These mathematical considerations, in
combination with naturalists' observations that widely aberrant indi;
uals were rarely found in natural populations (and when found, were
less viable), discredited macromutational theories of evolution and
theories emphasizing mutation pressure as the major factor in evolut
Similarly the belief that the random elimination of genes from popul
tions was an important factor in evolution, published by A. L. and A
Hagedoorn in their 1921 book, The Relative Value of the Processes Ca
Evolution,[25] was shown by Fisher in 1922 and later by Haldane and Wr
to be very unlikely in all but exceedingly small populations.

Perhaps the most pervasive way the models served to eliminate c
peting theories was by showing that many factors, especially the inh
tance of acquired characters, were not logically essential for the e
untionary process. All the ideas of directed evolution, and there w
many, were dealt a severe blow. Most evolutionists, like most scien
in general, accept some variation of the argument from parsimony.
gist of this argument is that a scientist should accept (within reas
able limits) the simplest hypothesis compatible with observed eviden
and should not multiply hypotheses without good reason. Combined wi
laboratory genetics, and what little was known of the genetics of na
populations, the new mathematical theory of selection provided that
simplest hypothesis for the mechanism of evolution. Competing theor
with their added hypotheses, suffered.

Reinterpretation of Data

My third conclusion is that the mathematical models provided a
framework for reinterpreting data already gathered from nature. The
most immediate impact of the models came from their application to da
gathered by persons who had little conception of quantitative evoluti
One of the earliest examples was the independent analysis of previous
published reports by Haldane (in 1924) and Chetverikov (in 1926) of
selective advantage enjoyed by the melanic form of the moth "Amphida
betularia (later transferred to Biston) over its non-pigmented form :
the area around Manchester, England.[26] Both Haldane and Chetverikov
calculated that a melanic form was twice as likely to survive as the
non-melanic.

Fisher's most important early studies in applied evolution came
from an analysis of data others had gathered for different purposes.
The first data came from E. B. Ford, who had gone into the field to
test E. B. Poulton's thesis that greater variability should be found
female rather than male Lepidoptera. Fisher had already argued in hi

)22 paper[27] that the survival of a rare mutant in a population was a
tochastic process independent of selection, and he concluded that var-
ability should therefore be greater in large than in small populations,
s Charles Darwin had predicted for other reasons in Chapter II of the
rigin. When analysed by Fisher, Ford's data did reveal on an average
reater variability in large populations. Two papers resulted from this
ollaboration.[28.]

Fisher published two more studies analysing the data of others in
he 1930's. One was an analysis of the data of R. K. Nabours on the
rouse locusts Apotettix eurycephalus and Paratettix texanus.[29] To
uttress his views on the evolution of dominance, Fisher wanted to demon-
trate that heterozygotes were plentiful in wild populations. From
nalyses of Nabours' data in 1930 and 1939, Fisher concluded that his
ypothesis was correct, and that his theory of the evolution of domin-
nce was strengthened. The second study was his 1937 paper[30] on the
elation between variability and abundance, an analysis of measurements
f the eggs of British nesting birds made by F. C. R. Jourdain. The
nalysis of the data indicated again that larger species vary more than
smaller.

Other examples of the application of the mathematical models to
data are plentiful, such as Haldane's estimation of mutation rates in
humans from examination of medical records on haemophilia,[31] or Cyril
Diver's use of Wright's model of optimum population structure to interpret data he had gathered on the distribution of the land snail Cepaea.[32]
Instead of listing these cases, I will turn to an example of a different
kind.

Paleontologists in the 1930's were probably the evolutionists least
interested in the new genetics. Many theories of evolution were compatible with the paleontological record, and genetic breeding experiments could not be carried out on fossils. But by the late 1930's it
became clear that someone really ought to show the compatibility of the
new quantitative evolution with what was known of paleontology. George
Gaylord Simpson performed this experiment in compatibility in his 1944
book Tempo and Mode in Evolution. Simpson's long chapter on the deter-
minants of evolution relies heavily and explicitly upon the work of
Fisher, Haldane, and Wright. Here, as in Dobzhansky's Genetics and the
Origin of Species, the mathematics is greatly simplified, but the models
were used to set the whole framework of Simpson's argument. Paleontologists reading Simpson's book would find more "common sense" of evolutionary mechanism than abstract mathematical models. But the models
unquestionably lent scientific legitimacy to Simpson's reasoning.

In each of these studies described above, the data assumed a new importance to evolutionists when the mathematical models were applied to them. I do not argue that the data were good enough to justify th conclusions reached by those who applied models to them, or that the models corresponded closely to actual situations in natural populatic only that the data became obviously more significant to evolutionists by being placed into a new framework of quantitative analysis.

Mathematical Models and Field Research

My fourth conclusion is that mathematical geneticists and their models stimulated and guided field researches on natural populations identifying clearly some relevant parameters of the evolutionary proc and by elucidating possible structures and dynamics of natural popula tions, thus influencing experimental design and generating, directly indirectly, a large number of hypotheses testable in the field. The mathematical models undeniably had a significant impact upon field re search.

From the beginning, Chetverikov, Fisher, Haldane, and Wright sta explicitly the function they hoped their quantitative work would serve They wished to elucidate the possible paths of evolution, and they sai that only empirical research upon natural populations could provide in formation on the relative weights to be placed upon the coefficients their equations, or their hypotheses about, for example, the evolution of dominance. Thus Haldane argued in 1929 that the role of his mathe matical work in biology was to suggest possibilities "which can only decided by observation of Nature, and not by experiments alone."[33] Chetverikov, Fisher, and Wright agreed. All four were perfectly aware that their models were not quantitative accounts of evolution in Natu but were guides for the effective study of that subject.

Many evolutionists approached the mathematical models in this spir Dobzhansky, for example, in the first edition of Genetics and the Orig of Species, stated that

> The experimental work, that should test these mathematical deduc-
> tions is still in the future, and the data that are necessary for
> the determination of even the most important constants in this fi
> are wholly lacking. Nonetheless, the results of the mathematical
> work are highly important, since they have helped to state clearl
> the problems that must be attacked experimentally if progress is
> to be made.

Later in the book he added: "the manner of action of selection has b dealt with only theoretically, by means of mathematical analysis. The

esults of this theoretical work (Haldane, Fisher, Wright) are however
nvaluable as a guide for any future experimental attack on the pro-
lem."[34]

N. W. Timofeef-Ressovsky expressed a view similar to that of Dob-
hansky in an important article, "Mutations and Geographical Variation,"
n the 1940 book, <u>The New Systematics</u>. There he argues that the work
f Chetverikov, Fisher, Haldane, and Wright

is of the greatest importance, showing us the relative efficacy of
various evolutionary factors under the different conditions possible
within the populations. It does not, however, tell us anything about
the real conditions in nature, or the actual empirical values of the
coefficients of mutation, selection, or isolation. It is the task
of the immediate future to discover the order of magnitude of these
coefficients in free-living populations of different plants and animals; this should form the aim and content of an empirical population genetics.[35]

n general, the mathematical models were developed well before much data
n natural populations was available. That they were extremely useful
or organizing the study of natural populations should not be surprising.
pecific examples of the influence of the models on field research a-
ound.

The obvious place to begin is with the influence of Chetverikov upon
the genetic analysis of natural populations. Because Prof. Mark Adams
f the University of Pennsylvania has already devoted two papers to this
topic,[36] I will only mention Chetverikov's role. As a result of his
mathematical analysis, Chetverikov became convinced that much variability lay hidden in natural populations. Dubinin, Timofeef-Ressovsky,
Gershenson, and others followed this lead and began, along with Chetverikov, to find surprisingly large numbers of hidden recessives in the
natural populations they studied.

The influence of Fisher, Haldane, and Wright upon field researches
is complicated by two factors which should always be borne in mind.
First, their views changed significantly over time, and second, their
disagreements with each other greatly stimulated field research. Only
a few illustrative examples can be provided here.

A. The analysis of polymorphism

In his 1922 paper Fisher outlined the possibility of a stable poly-
morphism in a population, perhaps caused by a balance of selection with
recurrent mutation, or by the greater fitness of the heterozygote as compared with the homozygote. But Fisher thought such stable polymorphisms

must be uncommon in nature, and he put little weight upon their impor
ance. Instead he emphasized the direct progressive slow action of se
lection upon single genes as the primary mechanisms of evolutionary
change. Soon, however, he found that polymorphisms were more importa:
than he had originally thought. In 1928 he published his theory of t:
evolution of dominance, which Sewall Wright quickly attacked because
the very low selection pressures, of the order of mutation pressures,
hypothesized by Fisher. The problem was that not enough heterozygote:
were available for more rapid selection. This problem could be alleviated in part if heterozygotes were more common in natural populations
than Fisher had earlier anticipated. Soon Fisher found evidence for a
stable over-abundance of heterozygotes in the work of Nabours on the
grouse locust. Suddenly Fisher became very interested in the stable
polymorphisms in nature.

Although by 1935 Fisher, Haldane, and Wright had all made room for
stable polymorphisms in their mathematical models, their primary atter
tion was to the progressive evolution of favorable genes or gene complexes. The problem which experimenters quickly encountered was that
cases like industrial melanism were rare, and there was little opportunity for field research upon them. The small selection pressures ir
progressive changes in gene frequencies were just too difficult to mea
sure. But if relatively permanent polymorphisms resulted from the gre
er fitness of the heterozygote, very large selection pressures against
the homozygotes might be held in equilibrium. These selection pressures could be measured. This is in my opinion the reason for the ger
eral shift in field studies observed in the 1930's from the study of
progressive selection of single genes to the study of polymorphic equi
ibria. What was so important about the mathematical models was that
room had been made for such polymorphisms at an early stage, and only
later was this possibility found to be crucial for field research.

The best known student of these stable polymorphisms is E. B. Ford
who with his students, has conducted a large number of field investiga
tions. His argument is that if polymorphisms exist because of a balan
of selective pressures against two alleles whose heterozygote is more
fit than either homozygote, then a change of conditions could unleash
these strong selective forces and produce relatively rapid evolutionar;
change.[37]

B. Dobzhansky's "Genetics of Natural Populations"

By far the most impressive series of papers which show the direct
influence of mathematical population geneticists is that on the "Gene-

ics of Natural Populations," begun in 1938 by Dobzhansky and his colleagues. Twenty-nine papers in this series had been published by 1959.
Dobzhansky's opinion concerning the influence of the mathematical models
upon field research has been quoted several times already. Sewall Wright
said in 1960 that "nobody ... has understood better than Dobzhansky the
reciprocal relation between concrete facts and abstract theory, as ex-
emplified in his continual collaboration with mathematical population
geneticists."[38] There is a special reason why Dobzhansky and Wright
could speak so authoritatively about collaboration: from 1937 until
1945 they corresponded voluminously, talked for many hours at scientific meetings, and arranged to visit each other as often as possible.
Dobzhansky quite clearly needed Wright's help in his studies of natural
populations. The only other comparable collaboration between a mathematical population geneticist and a naturalist was between Fisher and
Ford in England. Fortunately for them, but not for the historian of
science, Fisher and Ford talked rather than corresponded about sub-
stantive issues in evolutionary biology. The correspondence between
Wright and Dobzhansky is a unique and invaluable source of illumination
about the influence of mathematical population genetics upon field research in the 1930's and 40's. Only the briefest analysis of the correspondence is possible here.

Dobzhansky met Wright in 1932 at the Sixth International Congress
of Genetics in Ithaca. Wright delivered a seminal paper, containing
the first diagrammatic account of his famous fitness surfaces, with
their peaks and valleys. Dobzhansky was captivated, and as he wrote
later, "fell in love" with Wright at that meeting.[39] When Dobzhansky
was writing <u>Genetics and the Origin of Species</u> in 1937, he sent the
chapter on selection to Wright for comments. Wright replied in helpful
detail, and a fruitful collaboration began. They were always separated
by their jobs. Wright was at the University of Chicago, while Dobzhansky was at the California Institute of Technology and, after 1940, at
Columbia.

Dobzhansky had little mathematical training. He told me that he
regularly read Wright's papers by carefully reading the introductions,
rapidly skimming all sections with complex calculations, and carefully
reading the conclusions. Yet this method evidently enabled Dobzhansky
(and many other geneticists) to glean the essentials of Wright's views
on evolution.

Dobzhansky's first serious field researches after completing <u>Gene-</u>
<u>tics and the Origin of Species</u> concerned chromosomal and genic varia-
tion in natural populations of <u>Drosophila pseudoobscura</u>.[40] Sturtevant

had already analysed the frequency of lethals in the third chromosome
of wild populations of D. pseudoobscura, and Dobzhansky extended this
work. Both Dobzhansky and Sturtevant believed it was possible to estimate effective population size from comparisons of the rate of occu-
rence of lethals and their accumulation in the population. When Dobzhansky sent his preliminary results to Wright for comment, Wright replied that the estimation of effective population size by this method
would be difficult or impossible. Another exchange of letters pursued
the question in greater detail, with Wright's last response containing
some rather difficult-looking mathematics. Dobzhansky, who was proba-
unable to follow the exact reasoning, passed Wright's letter on to Stu-
tevant. Wright's complicated mathematics seemed totally unnecessary
Sturtevant, who wrote Wright a letter which began: "The calculation
seems to me to be extremely simple and easy. Have I made some fool s
or have I missed the point of what you and Dobzhansky are after?"[41]
There followed some very simple calculations purporting to clarify the
problem.

Wright was stirred to action by Sturtevant's letter, and he repl
in a letter addressed to both Sturtevant and Dobzhansky.[42] The lette
was eight single-spaced typewritten pages in length. It contained a
devastating critique of Sturtevant's letter, which in comparison appea
both naive and superficial. And Dobzhansky was less able to reason
quantitatively than Sturtevant! One result of this exchange was that
Dobzhansky realized how deeply he needed Wright's advice, and he came
to depend heavily upon Wright's help with experimental design and eval
uation of experimental data.

The full scale of this collaboration cannot be detailed here, but
it continued unabated until 1945, when Wright's other duties caused hi
to discourage further involvement. They published several papers join
where the collaboration was obvious. But the full extent of Dobzhansk
dependence upon Wright must be seen in their correspondence. One exam
will suffice. In January, 1943, Dobzhansky was invited to go to Brazi
for six months of teaching and research. He conceived a "grand plan"
of Drosophila research, which involved new experiments with Brazilian
species. But Dobzhansky did not want to leave without discussing the
plan with Wright, who might save him from costly mistakes of experimental design. Both men had heavy teaching schedules. Wright was una
to come to New York, so Dobzhansky wrote him: "I am reminded of the o
adage: 'If the mountain will not come to Mahomet, then Mahomet will g
to the mountain.'"[43] Dobzhansky took a train to Chicago, spent a day
talking to Wright, and caught the next train back to New York. There

n be no doubt that Wright's quantitative ideas about selection, pop-
ation structure, and random processes permeate Dobzhansky's work on
e genetics of natural populations. Later, of course, Dobzhansky de-
nded heavily upon statistician Howard Levene, whose role was similar
 that of Wright.

The stimulus of disagreement

A considerable amount of field research was stimulated by the dis-
reements between Fisher, Haldane, and Wright. The most prominent
ample was the controversy, between Fisher and Wright, over the struc-
re and relative size of populations in nature, and the role played by
ndom genetic drift in the production of novel heritable variation.
om the beginning of his work on evolution, Wright had emphasized that
 population could evolve much faster if it were divided up into small
cal populations between which some migration occurred and in which the
ndom fluctuations of gene frequencies could produce novel gene com-
nations. Fisher, on the other hand, at least through the time he wrote
e Genetical Theory of Natural Selection in 1930, was concerned with the
terministic effects of mass selection in very large populations. He
clared that laws governing natural selection were similar to the laws
 gases in physics. In the early 1930's, therefore, Fisher and Wright
sagreed sharply about the size of locally isolated populations one could
pect to find as a rule in nature, and about the relative roles of
lection and random processes. Dobzhansky's early papers on the "Gene-
cs of Natural Populations" series were directed primarily to the ques-
on of population size and structure. For the most part, Wright's
ews were vindicated by these studies.
But both Wright and Fisher were forced to partially change their
ews. Wright had suggested that because of random genetic drift, the
aracters distinquishing small local populations were likely to be of
ttle selective importance. For example, Wright argued in 1940 that
e chromosome polymorphisms found in wild populations of Drosophila
seudoobscura by Dobzhansky were probably of negligible selective im-
ortance.[44] But within a few years he and Dobzhansky discovered that
e polymorphisms changed drastically from season to season as a result
f powerful selection pressures on different chromosomal arrangements.[45]
isher had to change his views also. He began to realize that Wright
as correct in arguing that evolution would proceed more rapidly in a
opulation subdivided into partially isolated subpopulations. But he
till thought these small subpopulations were larger than those envis-
oned by Wright, and that differences between them were caused primarily

by selection pressures rather than Wright's genetic drift.

This issue came to direct confrontation in the 1940's. To suppo[rt] their view, Fisher and Ford studied between 1939 and 1946 the spread [of] a gene in an isolated colony of the moth _Panaxia dominula_, and compa[red] their findings with data about the colony gathered before 1929. The[ir] conclusions presented a direct challenge to Wright's view. They fou[nd]

> that the observed fluctuations in gene-ratio are much greater th[an] could be ascribed to random survival only. Fluctuations in natu[ral] selection (affecting large and small populations equally) must th[ere]fore be responsible for them. The possibility that random fluctu[a]tions in populations much smaller than 1000 could be of evolution[ary] importance is improbable in view of the frequency with which suc[h] small isolated populations must be terminated by extinction with[in] periods which must be extremely short from an evolutionary point [of] view.

> Thus our analysis, the first in which the relative parts playe[d] by random survival and selection in a wild population can be tes[ted] does not support the view that chance fluctuations in gene-ratio[s] such as may occur in very small isolated populations, can be of a[ny] significance in evolution.[46]

Wright responded vigorously in a 1948 article,[47] claiming that he nev[er] had placed such exclusive reliance upon random drift as Fisher and Fo[rd] imagined. He argued that the effective population size calculated b[y] Fisher and Ford might be too high, and that random drift might actua[lly] account for more of the fluctuation in gene frequency observed by the[m.] Fisher and Ford then answered Wright in a brief, sharply-worded artic[le.] The disagreement continued.

Other experimental work began to appear on this question. A. J. [Cain] and P. M. Sheppard analysed the relative importance of selection and random drift in populations of the snail _Cepaea nemoralis_. Diver ha[d] concluded in 1940 that the differences observed in small populations probably resulted from the random drift envisioned by Wright. Cain a[nd] Sheppard, following Fisher, tried to demonstrate by field studies tha[t] selection was more important. They concluded their investigation wit[h] the statement:

> Diver has claimed that variations in shell colour and banding in the snail _Cepaea nemoralis_ have no selective value, and occur at random in different colonies. He ascribes the differences betwee[n] colonies to genetical drift.

> These contentions cannot be sustained. There is a definite rel[a]tionship between the proportions of different varieties in any co[lony]

and the background on which they live ... there is good evidence that
the general appearance of any colony is determined by natural selection... All situations supposedly caused by drift should be rein-
vestigated.[49]

we saw in the opening section of this paper, Sheppard continued to
rk with Fisher's mathematical models as his guide, and he concentrated
imarily upon the conflicts between Fisher and Wright.

Enough examples have been cited in this section to demonstrate the
nsiderable direct influence the mathematical models had upon field
search. I have made no attempt here to assess carefully the indirect
fluence of the models upon field research, but I strongly suspect that
e distillation and exposition of the major features of the models in
rks like Dobzhansky's <u>Genetics and the Origin of Species</u>, of his papers
"Genetics of Natural Populations," had a significant impact upon
olutionary biologists.

Concluding Remarks

The mathematical models of theoretical population genetics were not
gical keys to the understanding of evolution. Every one of the models
ntained important simplifications, because otherwise the mathematical
lations were obviously too complex to handle. The data gathered from
ture were insufficient to "tune" the models finely. The models could
int out <u>possibilities</u>, but could not logically discriminate between
em. Although the models showed that in general evolution could pro-
ed by the action of surprisingly small selection pressures, and many
olutionists came to believe in this neo-Darwinian view, very little
perimental or field data supported this possible interpretation. Ob-
rvation of small selection pressures in nature is in practice an almost
possible task.

The mathematical models did, however, comprise one crucial part of
e evolutionary synthesis, as I have argued in this paper. In combin-
ion with advances in experimental genetics, and with knowledge of nat-
al populations often gathered by systematists with neo-Lamarckian
ews, the models exerted a considerable influence upon the views of
olutionists.

Very few evolutionists read the papers of Fisher, Haldane, and
ight with ease and full understanding of the mathematical analyses.
t they often selectively read the papers, paying close attention to
oblems addressed, assumptions, and conclusions. Fisher's 1918 paper
the correlation between relatives is difficult reading, but a math-
atical novice could understand that Fisher had tried to show that no

contradiction existed between Mendelism and observed correlations be
tween relatives in humans. Nor was it even necessary to read the ma
ematical papers to be influenced by them. Thus I believe some justi
cation exists for the almost automatic citation of the major works c
Fisher, Haldane, and Wright by many evolutionists during the 1930's
40's.

Notes

[1] I am greatly indebted to Ernst Mayr for his encouragement and h
and to Max Black and Dick Lewontin for a better understanding of mat
ematical models. The following persons contributed extremely helpfu
comments upon an earlier draft: I. Michael Lerner, Lewontin, Mayr,
Leigh Van Valen, and Sewall Wright. Part of the research was carrie
out under NSF grant #SOC 75-15367.

[2] Louis Trenchard More, Isaac Newton (New York: Dover, 1962), p.

[3] Theodosius Dobzhansky, "A Review of Some Fundamental Concepts a
Problems of Population Genetics," Cold Spring Harbor Symposia on Qu
titative Biology 20 (1955), p. 14.

[4] P. M. Sheppard, "Evolution in Bisexually Reproducing Organisms,
Evolution as a Process, ed. by Julian Huxley, A. C. Hardy, and E. B.
Ford (London: Allen and Unwin, 1953), pp. 201-202.

[5] C. H. Waddington, "Epigenetics and Evolution," Symposia of the
Society of Experimental Biology, Number VII (New York: Academic Pre
1953), p. 186.

[6] J. B. S. Haldane, "Foreword," ibid., p. ix.

[7] Ernst Mayr, "Where Are We?" Cold Spring Harbor Symposia on Quan
tative Biology 24 (1959), p. 2.

[8] Sewall Wright, "Genetics and Twentieth Century Darwinism: a re
and discussion," American Journal of Human Genetics 12 (1960), pp. 3
372.

[9] J. B. S. Haldane, "A Defense of Beanbag Genetics," Perspectives
Biology and Medicine 7 (1964), p. 344.

[10] Paul S. Moorhead and Martin M. Kaplan, eds., "Mathematical Chal
lenges to the Neo-Darwinian Interpretation of Evolution," The Wistar
Institute Symposium Monograph #5 (Philadelphia: Wistar Institute Pr
1967).

[11] I have written a book on this topic, and present only a very br
summary here. For discussion and detailed citations see William B.
vine, The Origins of Theoretical Population Genetics (Chicago: The
iversity of Chicago Press, 1971).

[12] David Starr Jordan and Vernon L. Kellogg, Evolution and Animal
(New York: D. Appleton, 1907), pp. 115-116.

[13] Theodosius Dobzhansky, Genetics and the Origin of Species (New
Columbia University Press, 1937. 2nd edition, 1941. 3rd edition, 1
Julian Huxley, Evolution: The Modern Synthesis (London: Allen and
win, 1942), and ed. The New Systematics (Oxford: Oxford University
Press, 1940); Ernst Mayr, Systematics and the Origin of Species (New
York: Columbia University Press, 1942); George Gaylord Simpson, Tem
and Mode in Evolution (New York: Columbia University Press, 1944);
Bernhard Rensch, Neuere Probleme der Abstammungslehre (Stuttgart: E

7), English translation of revised edition as <u>Evolution Above the</u>
<u>:ies Level</u> (New York: Columbia University Press, 1960).

[14] J. B. S. Haldane, "A Mathematical Theory of Natural and Artificial
ection," <u>Transactions of the Cambridge Philosophical Society 23</u>
24), p. 19.

[15] R. A. Fisher, "Retrospect of the Criticisms of the Theory of Nat-
l Selection," in <u>Evolution as a Process</u>, p. 85.

[16] Waddington, <u>Strategy of the Genes</u> (London: Allen and Unwin, 1957),
61.

[17] Mayr, "Where Are We?" p. 2.

[18] Charles Darwin, <u>On the Origin of Species</u> (facsimile of first edi-
n; Cambridge: Harvard University Press, 1966), pp. 186-187.

[19] Unpublished manuscript in possession of Ernst Mayr.

[20] E. B. Ford, <u>Mendelism and Evolution</u> (London: Methuen, 1931).

[21] Huxley, <u>The New Systematics</u>, pp. 2-3.

[22] J. B. S. Haldane, in "Evolution," <u>Symposia for the Society of Ex-</u>
<u>imental Biology 7</u> (New York: Academic Press, 1953), p. ix.

[23] Dobzhansky, <u>Genetics and the Origin of Species</u>, 2nd ed., p. 332.

[24] Gavin de Beer, ed., <u>Evolution</u> (Oxford: Oxford University Press,
8).

[25] A. L. and A. C. Hagedoorn, <u>The Relative Value of the Processes</u>
<u>sing Evolution</u> (The Hague: Martinus Nijhoff, 1921).

[26] Haldane, "A Mathematical Theory," p. 26; S. S. Chetverikov, "On
tain Aspects of the Evolutionary Process from the Standpoint of Mod-
Genetics" (trans. from the Russian by Malina Barker; ed. by I. Mich-
Lerner; originally published 1926), <u>Proceedings of the American Phil-</u>
<u>phical Society 105</u> (1961), pp. 184-185.

[27] R. A. Fisher, "On the Dominance Ratio," <u>Proceedings of the Royal</u>
<u>iety of Edinburgh 42</u> (1922), p. 324.

[28] R. A. Fisher and E. B. Ford, "Variability of Species", <u>Nature 118</u>
26), pp. 515-516; "The Variability of Species in the Lepidoptera, with
erence to Abundance and Sex", <u>Transactions of the Entomological Soc-</u>
y of London 76 (1929), pp. 367-384.

[29] R. A. Fisher, "The Evolution of Dominance in Certain Polymorphic
cies," <u>The American Naturalist 64</u> (1930), pp. 385-406; "Selective For-
in Wild Populations of <u>Paratetix texanus</u>", <u>Annals of Eugenics 9</u>
39), pp. 109-122. One object of a 1933 expedition by Nabours was to
her data for evaluating the theory presented by Fisher in the 1930 p
er cited above.

[30] R. A. Fisher, "The Relation Between Variability and Abundance Shown
the Measurements of the Eggs of British Nesting Bird," <u>Proceedings</u>
the Royal Society of London 122 B (1937), pp. 1-26.

[31] J. B. S. Haldane, <u>The Causes of Evolution</u> (New York: Harpers, 1932),
57.

[32] Cyril Diver, "The Problem of Closely Related Species Living in the
e Area," <u>The New Systematics</u>, pp. 323-328.

[33] J. B. S. Haldane, "The Species Problem in the Light of Genetics,"
ure 124 (1929), p. 516.

[34] Dobzhansky, <u>Genetics and the Origin of Species</u>, pp. 121, 176.

[35] N. W. Timofeef-Ressovsky, "Mutations and Geographical Variation,"
New Systematics, p. 104.

[36] Mark B. Adams, "The Founding of Population Genetics: Contribu
of the Chetverikov School," Journal of the History of Biology 1 (19
pp. 23-39; "Towards a Synthesis: Population Concepts in Russian Ev
tionary Thought", Journal of the History of Biology 3 (1970), pp. 1

[37] See Ford's Ecological Genetics (London: Methuen, 1964; 4th ed
1975) for a summary.

[38] Sewall Wright, "Genetics and Twentieth Century Darwinism," p.

[39] Undated letter from Dobzhansky to Provine.

[40] Th. Dobzhansky and M. L. Queal, "Genetics of Natural Populatio
I. Chromosome variation in populations of Drosophila pseudoobscura
habiting isolated mountain ranges", Genetics 23, pp. 239-251. Ibid
"Genic variations", ibid., pp. 463-484.

[41] Sturtevant to Wright, 16 September 1938.

[42] Wright to Dobzhansky and Sturtevant, 6 October 1938.

[43] Dobzhansky to Wright, 26 February 1943.

[44] Sewall Wright, "The Statistical Consequences of Mendelian Here
in Relation to Speciation," The New Systematics, pp. 178-179.

[45] Sewall Wright and Theodosius Dobzhansky, "Genetics of Natural
ulations. XII. Experimental Reproduction of Some of the Changes Ca
by Natural Selection in Certain Populations of Drosophila Pseudoobscura," Genetics 31 (1946), pp. 125-156.

[46] R. A. Fisher and E. B. Ford, "The Spread of a Gene in Natural
ditions in a Colony of the Moth Panaxia Dominula," Heredity 1 (1947
p. 173.

[47] Sewall Wright, "On the Roles of Directed and Random Changes in
Frequency in the Genetics of Populations," Evolution 2 (1948), pp.27

[48] R. A. Fisher and E. B. Ford, "The 'Sewall Wright Effect'," Her
4 (1950), pp. 117-119.

[49] A. J. Cain and P. M. Sheppard, "Selection in the Polymorphic L
Snail Cepaea Nemoralis," Heredity 4 (1950), pp. 275-294.

The first paper describing a specific mathematical model and its
tribution to biological discovery is Professor Walter's review of
historical development of a model for enzyme catalysis. Enzymes,
ch appear at the sub-cellular level in the biological hierarchy,
alyze and control individual biochemical steps in metabolic pathways.
hematical models of isolated enzymes have led to discovery about the
hanisms of enzyme catalysis and about how enzymes and their metabolites
trol biochemical pathways.

CONTRIBUTIONS OF ENZYME MODELS

Charles Walter
Department of Chemical Engineering
University of Houston
Houston, Texas 77004

I. Introduction

The recognition of enzymes is as old as our knowledge of catal
Quantitative studies of the actions of enzymes began in the latter
of the eighteenth century. In 1835 J. J. Berzelius included biolog
reactions as examples of reactions that occur under the influence o
new force which he called "catalytic". After completing a long lis
reactions of this type, Berzelius suggested, with amazing insight,
eventually it would be found that all the substances in living orga
are produced under the influence of this force.

In 1878 Kühne coined the term "enzyme" (in yeast). It was aro
this time that the well-known controversy between Liebig, who refus
accept that ferments required living cells, and Pasteur, who mainta
that fermentation was indissociably linked to the life process of t
cell, was raging. In this context, it became very important to est
lish whether or not enzymic reactions possessed characteristics in
mon with chemical reactions in general. It was thought that if enz
reactions were in some way linked with the mystery of life itself,
might not follow the laws obeyed by other chemical reactions. It wa
to resolve this issue that the first mathematical models of enzymes
were constructed.

Since other chemical reactions were known to follow the Law of
Action, it was this principle that was used to construct these math
matical models of enzyme action. These models contributed significa

subsequent discovery about enzyme action. The latter part of the nine-
nth century witnessed a large number of early kinetic experiments with
ymes. Many of the results seemed contradictory. For example, some
the early workers concluded that sucrose inversion catalyzed by in-
tase followed the Mass Action Law, but others, using whole yeast cells
cell-free extracts, obtained results that were apparently not in ac-
d with the Law of Mass Action (as it was known then). In some cases
re was a constant rate of product formation (zero-order kinetics),
reas in others the rate decreased from its initial value, but dif-
ently from what would be predicted by first-order kinetics.

Until 1902, all the mechanisms proposed to explain the kinetic be-
ior of enzymes were derived on the premise that the enzyme acted
ely by being present. Up until this time, no mechanism had been sug-
ted wherein the enzyme actually participates chemically in the re-
ion by combining with its substrates. However, when Brown (1902)
firmed his earlier findings that the kinetics of cell-free invertase
olved a slower decrease in the reaction rate that was predicted by
first-order equation, that all moderate sucrose concentrations gave
same initial rate, but lower sucrose concentrations gave a signifi-
tly lower rate, he reasoned that the enzyme E combines with its sub-
ate, S to form a compound, X, which then decomposes into the product,
ith regeneration of the free enzyme:

$$S + E \xrightleftharpoons[k_{-1}]{k_1} X_1 \xrightleftharpoons[k_{-2}]{k_2} P + E \qquad\qquad \text{Scheme I}$$

moderate sucrose concentrations the enzyme would be saturated with
strate so there could be no increase in the rate of product formation
n the substrate concentration was raised, but at lower sucrose con-
trations the enzyme would not be saturated so the rate would increase
decrease) when the substrate concentration was raised (or lowered).

Brown's interpretation of these results (Scheme I) was an important
step toward the construction of a model of enzyme action.

The Quasi-equilibrium Assumption

Henri (1903), and ten years later, Michaelis and Menten (1913),
translated Brown's chemical scheme into a mathematical model. Using
Law of Mass Action, it is possible to write four differential equati
relating the four dependent variables in Scheme I; when the system i
closed, it is possible to use the Law of Mass Conservation to write
additional equations relating these variables. Since the latter two
equations are algebraic, it is convenient to use them together with
of the differential equations for the model:

$$\dot{X}_1(t) = k_1 E(t)S(t) + k_{-2}E(t)P(t) - (k_{-1} + k_2)X_1(t)$$
$$\dot{P}(t) = k_2 X_1(t) - k_{-2}E(t)P(t)$$
$$S(0) = S(t) + X_1(t) + P(t)$$
$$E(0) = E(t) + X_1(t)$$

Equations (1)-(4) have not been solved analytically; Henri made
following simplfying assumptions:

(1) Assume that the enzyme, substrate, and enzyme-substrate c
     pound are in quasi-equilibrium. Then we can replace equat
     (1) with:

$$k_1 E(t)S(t) \approx k_{-1}X_1(t)$$

(2) Assume that the second term on the right hand side of equa
     (2) is negligible compared to the first one. Then we can

replace equation (2) with:

$$\dot{P}(t) \approx k_2 X_1(t) \tag{2a}$$

(3) Assume that the amount of substrate bound to the enzyme is small compared to the total amount of substrate initially present. Then we can replace equation (3) with:

$$S(0) \approx S(t) + P(t) \tag{3a}$$

Next, Henri used equations (3a) and (4) to eliminate $E(t)$ and $S(t)$ in equation (1a):

$$k_1 [E(0) - X_1(t)] [S(0) - P(t)] = k_{-1} X_1(t) \tag{1b}$$

solving equation (1b) for $X_1(t)$ and inserting the result into equation (2a), one obtains:

$$\dot{P}(t) = \frac{k_2 E(0)}{1 + \dfrac{k_{-1}}{k_1 [S(0) - P(t)]}} \tag{5}$$

Since, depending upon the magnitude of k_{-2}, the practical effect of assumption (2) can be to limit the applicability of the model to the early stages of the overall reaction, it is usual to assume in addition to (3) that:

(3a) the amount of product formed is small compared to the total amount of substrate initially present. Then we can replace equation (3a) with:

$$S(0) \approx S(t) \tag{3b}$$

Using equation (3b) in place of (3a), we obtain

$$v_e(0) \equiv \dot{P}(0) = \frac{V_m}{1 + \dfrac{K_s}{S(0)}} \tag{(}$$

where $V_m \equiv k_2 E(0)$ is the "maximum velocity" of the enzyme-catalyzed reaction, $K_s = k_{-1}/k_1$ is the dissociation constant for the enzyme-sub strate compound into enzyme and substrate, and $v_e(0)$ is the "initial, quasi-equilibrium rate". Equation (5a), often credited to Michaelis Menten (1913) was first derived by Henri (1903) ten years earlier.

The Quasi-steady State Assumption

Around this same time, Bodenstein (1913) pointed out that a less restrictive assumption than the quasi-equilibrium assumption is to su pose that the concentration of the enzyme-substrate compound is rela- tively steady in time compared to the rates of change of the substrate and product, but not necessarily in quasi-equilibrium with these reac- ants. This quasi-steady state assumption, treated more fully twelve years later by Briggs and Haldane (1925) permits us to replace equatic (1) with:

$$k_1 E(t)S(t) \approx (k_{-1} + k_2)X_1(t) \tag{(}$$

Note that the quasi-steady state equation, (1b) reduces to the quasi- equilibrium equation, (1a) when $k_{-1} \gg k_2$.

Briggs and Haldane (1925) used equation (1b) (rather than equatic (1a)) and equations (2a), (3b) and (4) to derive the model for the "ir itial, quasi-steady state rate":

$$v_{ss}(0) \equiv \dot{P}(0) = \frac{V_m}{1 + \frac{K_m}{S(0)}} \tag{5b}$$

where V_m has the same meaning as in equation (5a), but $K_m = (k_{-1} + k_2)$
$/k_1$ is generally not the dissociation constant for the enzyme-substrate
compound.

I. Validity of the Quasi-steady State Approximation

Equations (1)-(4) are the exact (insofar as the Laws of Mass Action
and Conservation are valid) mathematical description of the chemical
model in Scheme I. We may rewrite equations (1)-(4) in the more convenient form,

$$M(\dot{x},x) = 0 \tag{6}$$

where x is a vector in 4-space whose elements are the variables of the
system. In equation (6) and what follows chemical concentrations are
represented by the symbol for each component. A function, x* that sat-
isfies equation (6) is called an exact solution. The exact analytical
solution has not been obtained for equations (1)-(4).

It is not uncommon to find that the mathematical representation of
a model is facilitated if one makes an approximation. Such approximations introduce errors into the original model in the sense that the
approximate mathematical representation,

$$M(\dot{z},z) < e \tag{7}$$

is no longer an exact description of the actual model. Since (7) is an
inequality, there is not a single solution, but there is a range of z
which fulfills the conditions imposed by the magnitude of e. Since e

should be small and positive, the representation is approximately correct, and we refer to it as being ϵ-valid.

Henri and Briggs and Haldane were well-aware that equations (1)-(could not be solved analytically, and that, to express Brown's mechani as integrals of the Mass Action Law differential equations, it would b necessary to represent Scheme I by an approximate, rather than an exac mathematical model. It was for this reason - not because of any gener principle that indicated that the assumption would be a good approxima tion - that Henri and Briggs and Haldane used approximations to derive equations (5a) and (5b). It is therefore essential that we understand the implications of applying these assumptions to enzyme models.

Since the quasi-equilibrium assumption is a special case of the quasi-steady state assumption, we shall consider the implications of applying the latter to enzyme mechanisms. This approximation is valid if ϵ in inequality (7) is sufficiently small (i.e., if the quasi-steady state approximation does not introduce more than a maximum acceptable error into the model). Wong (1965) obtained an estimate of the maximu error introduced by applying the quasi-steady state approximation to Scheme I when $k_{-2} = 0$, and Walter (1974a) derived these maximum error estimates for the general situation when $k_{-2} \geq 0$. Using equation (4) to eliminate $E(t)$ from equation (1), one obtains

$$\frac{\dot{X}_1(t)}{k_1 S(t) + k_{-2} P(t) + k_{-1} + k_2} + X_1(t) = \frac{k_1 S(t) + k_{-2} P(t)}{k_1 S(t) + k_{-2} P(t) + k_{-1} + k_2} E(0) \tag{8}$$

The quasi-steady state estimates for $X_1(\text{ss})$ and $\dot{X}_1(\text{ss})$ are

$$X_1(\text{ss}) = \frac{k_1 S(\text{ss}) + k_{-2} P(\text{ss})}{k_1 S(\text{ss}) + k_{-2} P(\text{ss}) + k_{-1} + k_2} E(0) \tag{9}$$

$$\dot{X}_1(ss) = \frac{(k_1 - k_{-2})(k_{-1} + k_2)\dot{S}(ss)E(0)}{[k_1 S(ss) + k_{-2}P(ss) + k_{-1} + k_2]^2} \qquad (10)$$

If we substitute these quasi-steady state estimates into equation (8), we obtain

$$1 + \varepsilon \approx 1 \qquad (11)$$

where ε is the error introduced by the use of the quasi-steady state approximation:

$$\varepsilon = \frac{(k_1 - k_{-2})(k_{-1} + k_2)\dot{S}(ss)E(0)}{[k_1 S(ss) + k_{-2}P(ss)]\,[k_1 S(ss) + k_{-2}P(ss) + k_{-1} + k_2]^2} \qquad (12)$$

The quasi-steady state estimate for $\dot{S}(ss)$ is

$$-\dot{S}(ss) = \frac{\left[\dfrac{k_2 S(ss)}{K_f} - \dfrac{k_{-1}P(ss)}{K_r}\right]E(0)}{1 + \dfrac{S(ss)}{K_f} + \dfrac{P(ss)}{K_r}} \qquad (13)$$

Substitution into equation (12) yields

$$\varepsilon = \frac{\dfrac{k_2}{k_2 + k_{-1}}\dfrac{S(ss)}{K_f} - \dfrac{k_{-1}}{k_2 + k_{-1}}\dfrac{P(ss)}{K_r}}{\left[1 + \dfrac{S(ss)}{K_f} + \dfrac{P(ss)}{K_r}\right]^3}\left[\frac{(k_{-2} - k_1)}{k_1 S(ss) + k_{-2}P(ss)}\right]E(0) \qquad (14)$$

When $k_1 = k_{-2}$, $\varepsilon = 0$, and the quasi-steady state approximation has maximum validity (Walter, 1974a).

When $k_1 > k_{-2}$, $\varepsilon < 0$, and according to equation (14), the maximum value for $-\varepsilon$ occurs when k_{-2} is zero:

$$-\varepsilon < \frac{\frac{k_2}{k_2 + k_{-1}} \frac{S(ss)}{K_f}}{\left[1 + \frac{S(ss)}{K_f}\right]^3} \frac{E(0)}{S(ss)} \tag{1}$$

Since $0 < k_2/(k_2 + k_{-1}) < 1$,

$$-\varepsilon < \frac{\frac{S(ss)}{K_f}}{\left[1 + \frac{S(ss)}{K_f}\right]^3} \frac{E(0)}{S(ss)} \tag{1}$$

For any given $E(0)/S(ss)$ the value of $S(ss)/K_f$ giving the maximum possible value of $-\varepsilon$ can be calculated from

$$\frac{\partial(-\varepsilon)}{\partial \frac{S(ss)}{K_f}} = \frac{1 - 2 \frac{S(ss)}{K_f}}{\left[1 + \frac{S(ss)}{K_f}\right]^4} \frac{E(0)}{S(ss)} \tag{1}$$

by setting equation (17) equal to zero and solving for $S(ss)/K_f$:

$$\frac{S(ss)}{K_f} = \frac{1}{2} \tag{1}$$

Thus, when $k_1 \geq k_{-2}$ the maximum value possible for $-\varepsilon$ is

$$\varepsilon_{max} = \frac{4}{27} \frac{E(0)}{S(ss)} \qquad\qquad k_1 \geq k_{-2} \tag{1}$$

According to equation (19) the maximum error possible due to the quasi-steady state approximation applied to Scheme I when $k_1 \geq k_{-2}$ is $4/27$ of the $E(0)/S(ss)$ ratio (Wong, 1965; Walter, 1974a). A maximum error of one percent is expected when $E(0)/S(ss)$ is about $1/15$; a proportion-

tely smaller error is anticipated for smaller ratios of enzyme to sub-
trate.

When $k_1 < k_{-2}$, $\varepsilon > 0$, and the maximum value possible for ε is

$$\varepsilon_{max} = \frac{4}{27} \frac{E(0)}{S(ss)} \frac{k_{-2}}{k_1} \qquad\qquad k_1 < k_{-2} \qquad (20)$$

ccording to equation (20) the maximum error due to applying the quasi-
steady state approximation to the mechanism described in Scheme I when
$_1 < k_{-2}$ is 4/27 of the E(0)/S(ss) ratio _times_ the ratio of k_{-2}/k_1 (Wal-
er, 1974a). If $k_{-2} = 5k_1$, a maximum error of one percent is expected
when E(0)/S(ss) is about 1/75. However if $k_{-2} = 1000k_1$, E(0)/S(ss) would
have to be in the range of 15,000 to be sure that an error of one percent
was not introduced with the quasi-steady state approximation. Note that
f $k_{-2} = 1000k_1$ and $k_2 = 1000k_{-1}$, the equilibrium constant for the over-
ll reaction would be unity, and one would not know a _priori_ that k_{-2} was
so much larger than k_1.

Approximate Quasi-steady State Solutions During Time Intervals

Since enzymes are catalysts, they are usually employed in in vitro
experiments at lower concentrations than their substrates. Since chemical concentrations cannot be negative, the concentration range for the
various forms of the enzyme is

$$0 \leq X \leq E(0) \qquad (21)$$

On the other hand the range for the substrate or product is

$$0 \leq S \leq S(0) \qquad (22)$$

The range in inequality (22) is considerably larger than in inequality (21) whenever

$$S(0) \gg E(0) \qquad (2$$

Under these circumstances it is sometimes possible to treat the concentrations of the forms of the enzyme as quasi-constants over a time interval, t_1 to t_2.

To mathematicians, this argument for the quasi-steady state assumption seems scandalous at first. Strictly speaking, if X_1 in Scheme I is steady,

$$\dot{X}_1 = 0. \qquad (2$$

From the definition of thermodynamic equilibrium, we know that

$$\underset{t \to \infty}{\text{LIMIT}} \ \dot{X}(t) = 0. \qquad (2$$

But since we are interested in the nonequilibrium rates during the time interval t_1 to t_2 which is often taken early during the overall reaction the steady value of X_1 reached at thermodynamic equilibrium is not of general use to us here. Furthermore, since the reaction is initiated by mixing S and E, $P(0) = X_1(0) = 0$, and

$$\dot{X}_1(0) = k_1 E(0) S(0) \qquad (2$$

which, in general, is quite large. Evidentally then $\dot{X}_1(t)$ is initially positive and large and eventually small. If it does not change sign inbetween, equation (24) can never be true, and the steady state assumption cannot be strictly correct (Morales and Goldman, 1955). If we ifferentiate equation (1) with respect to time, we obtain

$$\ddot{X}_1(t) = - (k_1 S(t) + k_{-2} P(t) + k_{-1} + k_2) \dot{X}_1(t) +$$
$$(k_1 \dot{S}(t) + k_{-2} \dot{P}(t)) E(t).$$

(27)

ifferentiating equation (3), we see that at an instant, t* when equa-
ion (24) is true (i.e. when $\dot{X}_1(t^*) = 0$), $\dot{S}(t^*) = - \dot{P}(t^*)$. Substituting
nese relationships into equation (27), we obtain

$$\ddot{X}_1(t^*) = (k_{-2} - k_1) \dot{P}(t^*) E(t^*)$$

(28)

ince E(t) is necessairly positive and $\dot{P}(t)$ non-negative (Morales and
oldman, 1955), the right hand side of equation (28) is negative only
f $k_1 > k_{-2}$. Thus $\dot{X}_1(t)$ can change sign from its original positive value
o a negative value only when k_1 in Scheme I exceeds k_{-2}. This means
hat the steady state assumption (equation (24)) can be strictly correct
nly when $k_1 > k_{-2}$. On the other hand if $k_1 \leq k_{-2}$, $X_1(t)$ cannot pass
hrough a maximum and the steady state assumption cannot be exactly
orrect.

Mathematicians are quite right to be suspicious of an argument
hat uses the smallness of a (non-negative) variable to assert something
bout the smallness of the absolute value of its time derivative. How-
ver, in the case of the applicability of the quasi-steady state assump-
ion to Scheme I, the validity of the approximation is related to the
mallness of E(0)/S(0) through the singular perturbation theory of dif-
erential equations. This fact has been used by others in connection
ith specialized versions of Scheme I ($k_{-1} = k_{-2} = 0$, Yang, 1954; $k_1 \approx$
$_{-2}$, Miller and Alberty, 1958; $k_{-2} = 0$, Wong, 1965; Heineken et al.,
967). In what follows it is used in connection with the general mech-
nism in Scheme I.

If we differentiate equation (3) and substitute equations (1) and

(2) into the result, we obtain

$$\dot{S}(t) = - k_1 E(t)S(t) + k_{-1}X_1(t)$$

We next use equations (3a) and (4) to eliminate $E(t)$ and $P(t)$ from equations (2) and (29):

$$\dot{S}(t) = - k_1 E(0)S(t) + [k_1 S(t) + k_{-1}]X_1(t)$$

$$\dot{X}_1(t) = k_{-2}E(0)S(0) + (k_1 - k_{-2})E(0)S(t) -$$
$$[k_{-2}S(0) + (k_1 - k_{-2})S(t) + k_{-1} + k_2]X_1(t)$$

Equations (30) and (31) can be written in the dimensionless form,

$$\frac{dy}{d\gamma} \equiv f(y,z) = - y + (y + \alpha)z$$

$$\mathcal{M} \frac{dz}{d\gamma} \equiv g(y,z) = r + (1 - r)y - [(1 - r)y + r + \beta]z$$

where $y = S(t)/S(0)$, $z = X_1(t)/E(0)$, $\gamma = k_1 E(0)t$, $\mathcal{M} = E(0)/S(0)$, $r = k_{-2}/k_1$, $\alpha = k_{-1}/k_1 S(0)$, and $\beta = (k_{-1} + k_2)/k_1 S(0)$. We now consider the root of equation (31a) when $\mathcal{M} = 0$:

$$z \equiv \emptyset(y) = \frac{r + (1 - r)y}{\beta + r + (1 - r)y}$$

A theorem due to Tikhonov (1952) asserts that if for a fixed point, y the solution of the equation,

$$\frac{dz}{d\sigma} = g(y^*, z) \tag{33}$$

nds to $\emptyset(y^*)$ as $\sigma \rightarrow \infty$, the solution of equations (30a) and (31a),

$$y = Y(\tau, \mu) \tag{34}$$

$$z = Z(\tau, \mu) \tag{35}$$

nd to the solution of the approximate (quasi-steady state) system,

$$y = \overline{Y}(\tau) \tag{34a}$$

$$z = \emptyset(\overline{Y}(\tau)) \tag{35a}$$

the sense that

$$\underset{\mu \rightarrow 0}{\text{LIMIT}} \ Y(\tau, \mu) = \overline{Y}(\tau) \text{ in } 0 \leq \tau \leq T \tag{34b}$$

$$\underset{\mu \rightarrow 0}{\text{LIMIT}} \ Z(\tau, \mu) = \emptyset(\overline{Y}(\tau)) \text{ in } 0 \leq \tau \leq T \tag{35b}$$

For Scheme I, equation (33) is

$$\frac{(1 - r)\beta \frac{dy}{d\sigma}}{[\beta + r + (1 - r)y]^2} = r + (1 - r)y - [\beta + r + (1 - r)y]z \tag{33a}$$

rom the definition of thermodynamic equilibrium, the left hand side of
quation (33a) tends toward zero as $\sigma \rightarrow \infty$. We may therefore use Tik-
onov's theorem to assert that for sufficiently small $E(0)/S(0)$, the
olution of equations (30) and (31) tend to the solution of equation

(30) with

$$X_1(t) = \frac{\frac{k_{-2}}{k_1} + (1 - \frac{k_{-2}}{k_1}) \frac{S(t)}{S(0)}}{\frac{K_m}{S(0)} + \frac{k_{-2}}{k_1} + (1 - \frac{k_{-2}}{k_1}) \frac{S(t)}{S(0)}} E(0)$$

Substituting equation (32a) into (30), we obtain

$$\dot{S}(t) = \frac{- k_2 E(0) \left[(1 + 1/K) \frac{S(t)}{S(0)} - 1/K\right]}{\frac{K_m}{S(0)} + \frac{k_{-2}}{k_1} + (1 - \frac{k_{-2}}{k_1}) \frac{S(t)}{S(0)}}$$

where $K = k_1 k_2 / k_{-1} k_{-2}$, the equilibrium constant for the overall reac-
in Scheme I. Equation (34) is easily integrated

$$t = \frac{1 - \frac{k_{-2}}{k_1}}{V_m(1 + \frac{1}{K})} [S(0) - S(t)] - \frac{K_m + \frac{k_{-2}}{k_1} S(0) + \frac{1 - \frac{k_{-2}}{k_1}}{1 + K} S(0)}{V_m(1 + \frac{1}{K})} x$$

$$x \ln \frac{S(t) - S(\infty)}{S(0) - S(\infty)}$$

Equation (35) is the quasi-steady state solution of equation (2). Ti
honov's theorem guarantees the validity of equation (35) provided E(
S(0) is sufficiently small.

Claims such as those of Heinekin et al. (1967) that biochemists
can expect the quasi-steady state assumption to be valid provided E(
S(0) is "small" are somewhat misleading because how small E(0)/S(0)
be depends on the magnitude of $(k_1 - k_{-2})/k_1$ in Scheme I. We have s
(Equation (19)) that E(0)/S(0) << 1 is a necessary and sufficient co
dition for valid application of the quasi-steady state assumption wh

- k_{-2}▮/$k_1 \leq 1$, but when $k_1 \ll k_{-2}$, $E(0)/S(0) \ll 1$ no longer suffices

uation (20)). Thus Tikhonov's theorem assures the validity of equa-

on (35) when $E(0)/S(0)$ is sufficiently small and: 1) k_{-2} is not large

npared to k_1, or 2) for a given ratio of $k_{-2}/k_1 \gg 1$.

actical Effects of the Quasi-steady State Assumption

It is clear that when $k_{-2} > k_1$, the validity of the quasi-steady

ate approximation for the enzyme model in Scheme I depends on $\dot{X}_1$ being

all compared to $\dot{P}$ __after__ $P(t)$ has passed its inflection. Figure 1,

ken from the papers by Walter and Morales (1964) and Walter (1966),

lustrate the relationship between the ratio of $\dot{X}_1$ to $\dot{P}$ and the indi-

dual rate constants (Figure 1A) or $S(0)/E(0)$ (Figure 1B) in this model.

te that $\dot{X}_1$ can be 50 percent of $\dot{P}$ if k_{-2} is considerably greater than

or $S(0)$ is not sufficiently large compared to $E(0)$.

Figures 2, 3, and 4, taken from the paper by Walter (1966), illus-

ate the effect of steady state invalidity on plots of $\dot{P}$ versus $E(0)$

igure 2), $\dot{P}$ versus $\dot{P}/S(0)$ (Figure 3), and the values of the Michaelis

nstant (K_f) and the maximum velocity (V_f) calculated from a nonlinear,

ast square fit of $\dot{P}$ and $S(0)$ to the usual rectangular hyperbola (Fig-

e 4). It is clear from these figures that both K_f and V_f can be in

ror by 50 to 100 percent when k_{-2} is large compared to k_1 or $S(0)$ is

t sufficiently larger than $E(0)$.

I. An Alternative Approximation: Inflection Point Enzyme Kinetics

For many enzyme mechanisms, including Scheme I, $P(t)$ experiences

inflection. At the instant of inflection,

$$\ddot{P}(t^*) = 0 \tag{36}$$

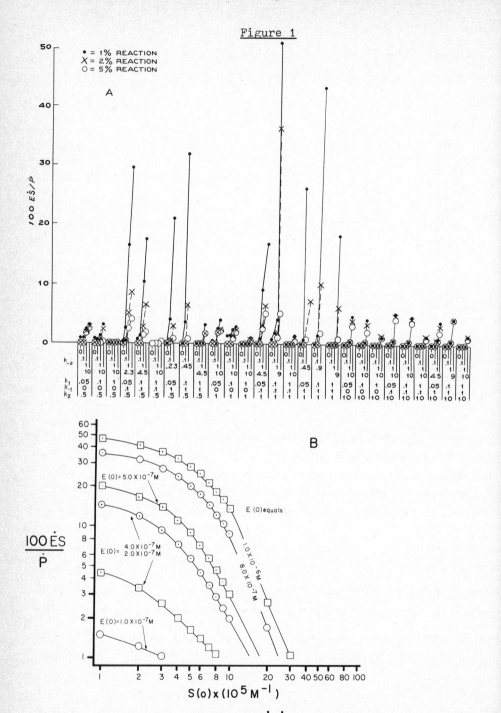

Quasi-steady state error ($100E\dot{S}/\dot{P}$) arising in Scheme I for diffe
values of the individual rate constants (Figure 1A) and different ini
conditions (Figure 1B). In Figure 1A $S(0) = 100E(0) = 0.1$ mM; in Fig
1B $k_1 = 20/mM\tau$, $k_{-1} = .1/\tau$, $k_2 = 1.9/\tau$, and $k_{-2} = 380/mM\tau$ where τ is
time unit for the rate constants.

Figure 2

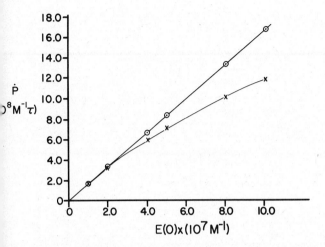

Relationship between $\dot{P}$ and $E(0)$ for the mechanism in Scheme I. The
.lues of the rate constants are the same as those in Figure 1B. O—O
; the quasi-steady state plot, and X—X is the plot obtained from
.uations (1)-(4) when no approximations are made.

.nce the instant $P(t)$ inflects can be determined from $P(t)$ itself, equa-
.on (36) provides a basis for the simplification of the differential
.uations describing enzyme mechanisms which, unlike the quasi-steady
:ate assumption, does not require information about the magnitude of
.(t) or its derivative (Walter, 1966).

Using equations (3) and (4) to eleminate $S(t)$ and $E(t)$ from equa-
.ons (1) and (2), we obtain

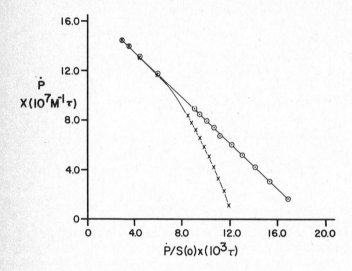

Relationship between $\dot{P}$ and $\dot{P}/S(0)$ for the mechanism in Scheme I. The values of the rate constants are identical to those in Figure 1B. O—O is the quasi-steady state plot, and X——X is the plot obtained from equations (1)-(4) when no approximations are made.

$$\dot{X}_1(t) = k_1[S(0) - X_1(t) - P(t)][E(0) - X_1(t)] +$$

$$k_{-2}P(t)[E(0) - X_1(t)] - (k_{-1} + k_2)X_1(t)$$

$$\dot{P}(t) = k_2X_1(t) - k_{-2}P(t)[E(0) - X_1(t)]$$

Differentiating equation (38) with respect to time and introducing equation (36) (inflection point kinetics) into the result, we obtain a relationship among $\dot{X}_1(t^*)$, $X_1(t^*)$, and $P(t^*)$:

Figure 4

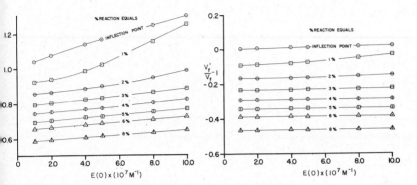

Relationship between calculated values of K_f (K_f' in Figure 4A) and V_f (V_f' in Figure 4B), the percent reaction at which P was estimated, and E(0) for the mechanism in Scheme I. The values of the rate constants are the same as in Figure 1B.

$$\ddot{P}(t*) = 0 = k_2\dot{X}_1(t*) - k_{-2}\Big[[E(0) - X_1(t*)][k_2X_1(t*) - k_{-2}P(t*)[E(0) - X_1(t*)]\Big] + P(t*)\dot{X}_1(t*)\Big]$$
(39)

Using equation (39) to eliminate $\dot{X}_1$ from equation (37), we obtain

$$[k_2 + k_{-2}P(t^*)]\Big[k_1[S(0) - X_1(t^*) - P(t^*)][E(0) - X_1(t^*)] +$$

$$k_{-2}P(t^*)[E(0) - X_1(t^*)] - (k_{-1} + k_2)X_1(t^*)\Big] = \qquad ($$

$$k_{-2}[E(0) - X_1(t^*)][k_2X_1(t^*) - k_{-2}P(t^*)[E(0) - X_1(t^*)]\Big]$$

Equation (40) provides a relationship between $P(t^*)$ and $X_1(t^*)$ which
be used to eleminate X_1 from equation (38) (Walter, 1966). The resul
ing expression relates the inflection point rate, $\dot{P}(t^*)$ and its integral, $P(t^*)$ to the initial conditions and rate constants in Scheme I.
An exact solution is thus obtained for the mathematical model describ.
Scheme I when $P(t)$ is observed in the vicinity of its inflection poin

$$P(t_1) < P(t^*) < P(t_2) \qquad ($$

IV. <u>Quasi-steady State Enzyme Kinetics: Manual and Computer-based
Derivations of Rate Equations</u>

<u>Equations Relating Quasi-steady State Rates and Initial Substrate Con-
centrations</u>

An important general method for deriving quasi-steady state rate
equations for enzyme mechanisms (King and Altman, 1956) utilizes Crame
role in a relatively uncomplicated way. For an enzyme mechanism involving n + 1 different enzyme forms (X_i, i = 0,1,...,n) including X_o
E, the concentration of each X_i relative to the total enzyme concentra
tion, $X_i/E(0)$ is a quotient of two summation terms. Each term is the
product of n different unimolecular rate constants or bimolecular rate
constants times a reactant concentration. Each term in the numerator
of $X_m/E(0)$ involves the rate constants (and the reactants for bimolecular constants) associated with reaction steps that individually or
in sequence lead to X_m. The n rate constants in each term are associ-

ted with the n reaction steps in which each of the n different $X_i (i \neq m)$
s a reactant. All of the possible combinations of n rate constants that
onform to this statement are present in the numerator of $X_m/E(0)$. The
enominator is the sum of all the various numerators for $m = 0,1,\ldots,n$.

The following two examples should help to illustrate the advantages
f this schematic method of deriving quasi-steady state rate equations.
he mechanism involving two enzyme intermediates $(n = 2)$ is

$$X_o + S \underset{k_{-1}}{\overset{k_1}{\rightleftharpoons}} X_1 \underset{k_{-2}}{\overset{k_2}{\rightleftharpoons}} X_2 \underset{k_{-3}}{\overset{k_3}{\rightleftharpoons}} X_0 + P$$

Scheme II

he numerator and denominator terms can be obtained by inspecting Figure
5. This figure contains the direct and sequential steps that lead to
each X_m. The relative concentration of each enzyme-containing species
is proportional to the sum of three terms:

$$\frac{X_o}{E(0)} \propto k_{-1}k_3 + k_{-1}k_{-2} + k_2k_3 \tag{42}$$

$$\frac{X_1}{E(0)} \propto k_1 S k_{-2} + k_1 S k_3 + k_{-2}k_{-3}P \tag{43}$$

$$\frac{X_2}{E(0)} \propto k_2 k_{-3}P + k_1 S k_2 + k_{-1}k_{-3}P \tag{44}$$

The proportionality constant for equations (42)-(44) is $1/(X_o + X_1 + X_2)$.
From the quasi-steady state rate of product formation,

Figure 5

m	Direct Steps	Sequential Steps	
0	$\xleftarrow{k_{-1}}$ $\xrightarrow{k_3}$	$\xleftarrow{k_{-1}}$ $\xleftarrow{k_{-2}}$	$\xrightarrow{k_2}$ $\xrightarrow{k_3}$
1	$\xrightarrow{k_1 S}$ $\xleftarrow{k_{-2}}$	$\xrightarrow{k_1 S}$ $\xrightarrow{k_3}$	$\xleftarrow{k_{-2}}$ $\xleftarrow{k_{-3}P}$
2	$\xrightarrow{k_2}$ $\xleftarrow{k_{-3}P}$	$\xrightarrow{k_1 S}$ $\xrightarrow{k_2}$	$\xleftarrow{k_{-1}}$ $\xleftarrow{k_{-3}P}$

Direct and Sequential Steps leading to X_0, X_1, and X_2 in the mechanism described by Scheme II in the text.

$$\dot{P} = \frac{k_3 X_2 - k_{-3} X_0 P}{X_0 + X_1 + X_2} E(0),$$

(4̲

we obtain the quasi-steady state rate equation,

$$\dot{P} = \frac{(k_1 k_2 k_3 S - k_{-1} k_{-2} k_{-3} P) E(0)}{k_{-1} k_3 + k_{-1} k_{-2} + k_2 k_3 + k_1 (k_{-2} + k_3 + k_2) S + k_{-3}(k_{-2} + k_{-1} + k_2) P}$$

(46

The mechanism involving three enzyme intermediates ($n = 3$) is

$$S + X_0 \underset{k_{-1}}{\overset{k_1}{\rightleftharpoons}} X_1 \underset{k_{-2}}{\overset{k_2}{\rightleftharpoons}} X_2 \underset{k_{-3}}{\overset{k_3}{\rightleftharpoons}} X_3 \underset{k_{-4}}{\overset{k_4}{\rightleftharpoons}} X_0 + P$$

Scheme III

he direct and sequential steps for each of the X_m in Scheme III appear n Figure 6. Inspection of this figure reveals the relative concentra- ions for each enzyme-containing compound:

Figure 6

m	Direct Steps		Sequential Steps	
0	$\xleftarrow{k_{-1}}$ $\quad \xrightarrow{k_3}\xrightarrow{k_4}$	$\xleftarrow{k_{-1}}\xleftarrow{k_{-2}}$ $\quad \xrightarrow{k_4}$	$\xrightarrow{k_2}\xrightarrow{k_3}\xrightarrow{k_4}$	$\xleftarrow{k_{-1}}\xleftarrow{k_{-2}}\xleftarrow{k_{-3}}$
1	$\xrightarrow{k_1 S}$ $\quad \xleftarrow{k_{-2}}\xleftarrow{k_{-3}}$	$\xrightarrow{k_1 S}$ $\xleftarrow{k_{-2}}$ $\quad \xrightarrow{k_4}$	$\xrightarrow{k_1 S}$ $\quad \xrightarrow{k_3}\xrightarrow{k_4}$	$\xleftarrow{k_{-2}}\xleftarrow{k_{-3}}\xleftarrow{k_{-4}P}$
2	$\xrightarrow{k_2}$ $\quad \xleftarrow{k_{-3}}\xleftarrow{k_{-4}P}$	$\xrightarrow{k_1 S}\xrightarrow{k_2}$ $\quad \xleftarrow{k_{-3}}$	$\xrightarrow{k_1 S}\xrightarrow{k_2}$ $\quad \xrightarrow{k_4}$	$\xleftarrow{k_{-1}}$ $\quad \xleftarrow{k_{-3}}\xleftarrow{k_{-4}P}$
3	$\xleftarrow{k_{-1}}$ $\xrightarrow{k_3}$ $\xleftarrow{k_{-4}P}$	$\xrightarrow{k_2}\xrightarrow{k_3}$ $\xleftarrow{k_{-4}P}$	$\xrightarrow{k_1 S}\xrightarrow{k_2}\xrightarrow{k_3}$	$\xleftarrow{k_{-1}}\xleftarrow{k_{-2}}$ $\quad \xleftarrow{k_{-4}P}$

Direct and Sequential Steps leading to $X_0, X_1, X_2,$ and X_3 in the mecha- nism described by Scheme III in the text.

$$\frac{X_o}{E(0)} \propto k_{-1}k_{-2}k_{-3} + k_{-1}k_{-2}k_4 + k_{-1}k_3k_4 + k_2k_3k_4 \tag{(}$$

$$\frac{X_1}{E(0)} \propto k_1Sk_{-2}k_{-3} + k_1Sk_{-2}k_4 + k_1Sk_3k_4 + k_{-2}k_{-3}k_{-4}P \tag{(}$$

$$\frac{X_2}{E(0)} \propto k_1Sk_2k_{-3} + k_1Sk_2k_4 + k_{-1}k_{-3}k_{-4}P + k_2k_{-3}k_{-4}P \tag{(}$$

$$\frac{X_3}{E(0)} \propto k_1Sk_2k_3 + k_{-1}k_{-2}k_{-4}P + k_{-1}k_3k_{-4}P + k_2k_3k_{-4}P \tag{(}$$

The proportionality constant for equations (47)-(50) is $1/(X_o + X_1 + $ $+ X_3)$. For this case the quasi-steady state rate of product formatio

$$\dot{P} = \frac{k_4X_3 - k_{-4}X_oP}{X_o + X_1 + X_2 + X_3} E(0) \tag{(}$$

The quasi-steady state rate equation for Scheme III is

$$\dot{P} = \frac{(k_1k_2k_3k_4S - k_{-1}k_{-2}k_{-3}k_{-4}P)E(0)}{\begin{array}{l} k_{-1}k_3k_4 + k_{-1}k_{-2}k_4 + k_2k_3k_4 + k_{-1}k_{-2}k_{-3} + k_1(k_{-2}k_{-3} + k_{-2}k_4 + k_3k_4 \\ + k_2k_{-3} + k_2k_4 + k_2k_3)S + k_{-4}(k_{-2}k_{-3} + k_{-1}k_{-3} + k_{-1}k_3 + k_2k_{-3} + \\ + k_{-1}k_{-2} + k_2k_3)P \end{array}} \tag{(.}$$

Computer-based Derviations of Quasi-steady State Rate Equations

It is evident from these relatively simple enzyme mechanisms tha
pencil and paper derivations of quasi-steady state rate equations are
tedious. However since the procedure for obtaining them is straightfo
ward, it is easily programmed for computer-based derivations. The im-
plementation of the King-Altman method into a digital computer progran

isher and Schultz, 1969; Schultz and Fisher, 1969) was a significant

velopment in quasi-steady state enzyme kinetics because it made all

bsequent manual derivations unnecessary. It is certainly no longer

cience" to spend several hours deriving algebraic expressions that a

mputer can generate <u>correctly</u> much more quickly.

The algorithm to determine quasi-steady state rate equations is

esented in Figure 7. The symbolism is that developed by Iverson (1962,

>63, 1964).

Figure 7

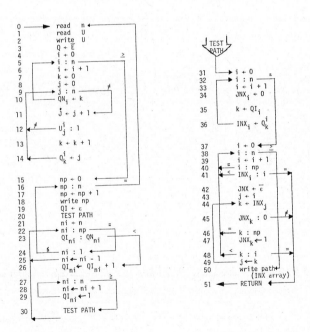

Program to determine rate equation coefficients.

Use of the program to derive quasi-steady state rates is also straightforward. The computer input consists of a "connection matri[x] the number of forward-direction reaction nodes terminating at the fr[ee] enzyme, and a description of the substrate-binding and product-desor[p] tion steps in the mechanism. The "connection matrix", U associated [with] a mechanism consists of elements, $^i u_j$ where $^i u_j = 1$ if there is a rea[c] tion path from source node i to destination node j, but $^i u_j = 0$ other[-] wise. In the connection matrix the numbers in the rows (R's) corres[pond] to the source nodes for each elementary chemical reaction, and the nu[m] bers in the columns (C's) to the destination nodes. For example.

$$
U \quad = \quad
\begin{array}{ccc}
C1 & C2 & C3 \\
\downarrow & \downarrow & \downarrow \\
0 & 1 & 1 \leftarrow R1 \\
1 & 0 & 1 \leftarrow R2 \\
1 & 1 & 0 \leftarrow R3
\end{array}
$$

is the connection matrix for the mechanism in Scheme II. The diagona[l] entries are zeros because there are no autocatalytic steps. C2-R1 re[fers] to the step wherein X_0 is converted into X_1 (the step described by k_1 C3-R1 to the X_0 to X_2 step (k_{-3}); C1-R2 to the X_1 to X_0 step (k_{-1}); C3-R2 to the X_1 to X_2 step (k_2); C1-R3 to the X_2 to X_0 step (k_3); C2-[R3] to the X_2 to X_1 step (k_{-2}). There are three forward-direction reacti[on] nodes terminating at the free enzyme, and

X0 S) P)
X1
X2

describes the first step as the substrate-binding step and the last s[tep] as the product-desorption step.

$$
\begin{array}{c}
\quad\;\; C1 \quad C2 \quad C3 \quad C4 \\
\quad\;\; \downarrow \quad\;\; \downarrow \quad\;\; \downarrow \quad\;\; \downarrow \\
U \; = \;
\begin{array}{cccc}
0 & 1 & 0 & 1 \;\leftarrow R1 \\
1 & 0 & 1 & 0 \;\leftarrow R2 \\
0 & 1 & 0 & 1 \;\leftarrow R3 \\
1 & 0 & 1 & 0 \;\leftarrow R4
\end{array}
\end{array}
$$

the connection matrix for the mechanism in Scheme III. As before, the

agonal entries are zeros because there are no autocatalytic steps.

wever in this connection matrix C3-R1, C4-R2, C1-R3, and C2-R4 also

ve zero entries because there are no chemical steps wherein X_0 is

nverted into X_2 (C3-R1), X_1 is converted into X_3 (C4-R2), X_2 is con-

rted into X_0 (C1-R3), or X_3 is converted into X_1 (C2-R4). In this

chanism there are four nodes terminating at the free enzyme, and

```
X0  S)      P)
X1
        X2
          X3
```

scribes the mechanism. If there was an additional substrate, R cap-

le of binding only X_1, and an additional product, Q desorbed in the

ep when X_3 is formed, the input description would be

```
X0  S)      P)
    X1  R)
    X2
    Q)  X3
```

The output from the quasi-steady state rate equation computer pro-

am consists of the "terminal node matrix" for each terminal node of

e reaction scheme and the coefficients that appear in the numerator

the rate equation. Each row of the terminal node matrix refers to

e group of rate constants that constitute a term in the denominator

the rate equation. The column numbers of this matrix refer to the

urce node for the step described by the rate constant, and if the entry

not zero, the numerical value refers to the destination node of the step. If the entry is zero, there is no reaction between the indica

steps for the mechanism input to the computer. For example, the ter▮

node matrix

$$
\begin{array}{c}
\;\; \overset{\text{C1}}{\downarrow}\;\; \overset{\text{C2}}{\downarrow}\;\; \overset{\text{C3}}{\downarrow} \\
\begin{array}{cccc}
+ & 0 & 1 & 1 \;\leftarrow\text{R1} \\
+ & 0 & 1 & 2 \;\leftarrow\text{R2} \\
+ & 0 & 3 & 1 \;\leftarrow\text{R3}
\end{array}
\end{array}
$$

is for terminal node 1 (the nodes terminating at X_o) for the mechani▮

described by Scheme II. The plus signs indicate positive coefficien▮

The entries in the first column are zeros because there are no autoc▮

alytic steps involving X_o. The 1 in C2-R1 indicates the X_1 to X_o (k▮

step, and the 1 in C3-R1 indicates the X_2 to X_o (k_3) step. The firs▮

row (R1) therefore indicates that $+k_{-1}k_3$ is one of the terms in the ▮

proportional to X_o. The 1 in C2-R2 again indicates the X_1 to X_o (k▮

step, and the 2 in C3-R2 indicates the X_2 to X_1 (k_{-2}) step. R2 ther▮

indicates that $+k_{-1}k_{-2}$ is one of the terms in the sum proportional t▮

X_o. The 3 in C2-R3 indicates the X_1 to X_2 step (k_2), and the 1 in C▮

indicates the X_2 to X_o step (k_3). Row 3 therefore indicates that $+k$▮

is one of the terms in the sum proportion to X_o. Terminal node matr▮

one therefore yields the three terms

$$+k_{-1}k_3 + k_{-1}k_{-2} + k_2k_3$$

which are the terms whose sum is proportional to X_o (See equation 42)

The terminal node matrix

$$
\begin{array}{c}
\;\; \overset{\text{C1}}{\downarrow}\;\; \overset{\text{C2}}{\downarrow}\;\; \overset{\text{C3}}{\downarrow} \\
\begin{array}{llcccc}
\text{S)} & + & 2 & 0 & 1 \;\leftarrow\text{R1} \\
\text{S)} & + & 2 & 0 & 2 \;\leftarrow\text{R2} \\
\text{P)} & + & 3 & 0 & 2 \;\leftarrow\text{R3}
\end{array}
\end{array}
$$

for terminal node 2 (the reaction nodes terminating at X_1) for Scheme

C1-R1 yields k_1, and C3-R1 k_3. Row 1 therefore indicates that $+Sk_1$

is one of the terms in the sum proportional to X_1. C1-R2 yields k_1,

₁ C3-R2 k_{-2}; R2 therefore indicates that $+Sk_1k_{-2}$ is one of the terms

the sum portional to X_1. C1-R3 yields k_{-3}, and C3-R3 k_{-2}; R3 there-

ʳe indicates that $+Pk_{-2}k_{-3}$ is one of the terms in the sum proportional

X_1. Terminal node matrix two therefore yields the three terms

$$+ Sk_1k_3 + Sk_1k_{-2} + Pk_{-2}k_{-3}$$

.ch are the terms whose sum is proportional to X_1 (See equation (43)).

The terminal node matrix

S)	+	2	3	0
P)	+	3	1	0
P)	+	3	3	0

for terminal node three (the nodes terminating at X_2) for Scheme II.

w one yields the term $+Sk_1k_2$, row two $+Pk_{-1}k_{-3}$, and row three $+Pk_2k_{-3}$.

is terminal node matrix therefore yields the terms

$$+ Sk_1k_2 + Pk_{-1}k_{-3} + Pk_2k_{-3}$$

ich are the terms whose sum is proportional to X_2 (See equation (44)).

The denominator of the quasi-steady state rate equation is the sum

all the terms generated by all <u>valid</u> terminal nodes:

$$_{-1}k_{-2} + k_{-1}k_3 + k_2k_3 + k_1(k_2 + k_{-2} + k_3)S + k_{-3}(k_{-1} + k_2 + k_{-2})P$$

ʳe equation (46)).

The rate equation coefficients generated by the computer program for this mechanism are

$$
\begin{array}{ccccc}
\text{S)} & + & 2 & 3 & 1 \\
\text{P)} & - & 3 & 1 & 2
\end{array}
$$

From the first row we obtain $+Sk_1k_2k_3$ and from the second, $-Pk_{-1}k_{-2}k$

The numerator for the quasi-steady state rate equation is therefore

$$+ \quad Sk_1k_2k_3 - Pk_{-1}k_{-2}k_{-3}$$

These combinations of rate constants for the numerator and the denom

nator yield equation (46), the quasi-steady state rate equation for

Scheme II.

The matrix for terminal node one of the mechanism described by

Scheme III is

$$
\begin{array}{ccccc}
+ & 0 & 1 & 2 & 1 \\
+ & 0 & 1 & 2 & 3 \\
+ & 0 & 1 & 4 & 1 \\
+ & 0 & 3 & 4 & 1
\end{array}
$$

This terminal node matrix yields the sum

$$+ \ k_{-1}k_{-2}k_4 + k_{-1}k_{-2}k_{-3} + k_{-1}k_3k_4 + k_2k_3k_4$$

which are the terms whose sum is proportional to X_o (See equation (4?)

Terminal node matrix two for Scheme III is

$$
\begin{array}{cccccc}
\text{S)} & + & 2 & 0 & 2 & 1 \\
\text{S)} & + & 2 & 0 & 2 & 3 \\
\text{S)} & + & 2 & 0 & 4 & 1 \\
\text{P)} & + & 4 & 0 & 2 & 3
\end{array}
$$

This terminal node matrix yields the sum

$$+ \ Sk_1k_{-2}k_4 + Sk_1k_{-2}k_{-3} + Sk_1k_3k_4 + Pk_{-4}k_{-2}k_{-3}$$

ich corresponds to the four terms proportional to X_1 in equation (48).
Terminal node matrix three for this mechanism is

$$
\begin{array}{llrrrr}
S) & + & 2 & 3 & 0 & 1 \\
S) & + & 2 & 3 & 0 & 3 \\
P) & + & 4 & 1 & 0 & 3 \\
P) & + & 4 & 3 & 0 & 3 \\
\end{array}
$$

is terminal node matrix yields the sum

$$+ \ Sk_1k_2k_4 + Sk_1k_2k_{-3} + Pk_{-4}k_{-1}k_{-3} + Pk_{-4}k_2k_{-3}$$

ich corresponds to the four terms proportional to X_2 in equation (49).
Terminal node matrix four for this mechanism is

$$
\begin{array}{llrrrr}
S) & + & 2 & 3 & 4 & 0 \\
P) & + & 4 & 1 & 2 & 0 \\
P) & + & 4 & 1 & 4 & 0 \\
P) & + & 4 & 3 & 4 & 0 \\
\end{array}
$$

his matrix yields the sum

$$+ \ Sk_1k_2k_3 + Pk_{-4}k_{-1}k_{-2} + Pk_{-4}k_{-1}k_2 + Pk_{-4}k_2k_3$$

ese terms correspond to the four terms proportional to X_3 in equation
0).

The rate equation coefficients generated for the mechanism are

$$
\begin{array}{llrrrr}
S) & + & 2 & 3 & 4 & 1 \\
P) & - & 4 & 1 & 2 & 3 \\
\end{array}
$$

hese yield the sum

$$+ Sk_1k_2k_3k_4 - Pk_{-4}k_{-1}k_{-2}k_{-3}$$

which corresponds to the numerator of the quasi-steady state rate eq
tion (See equation (52)). The sum of the sixteen terms generated by
four terminal node matrices correspond to the denominator of this eq
tion.

V. Acquisition and Analysis of Quasi-steady State Data.

Methods for Estimating Initial Rates

Quasi-steady state enzyme kinetic data usually consists of a se
of substrate or product concentrations, c_i obtained at discrete time
t_i, or of a continuous record of $c(t)$ plotted on a stripchart record
The usual procedure in enzyme studies is to use the kinetic data to
tain a single parameter called the "initial, quasi-steady state rate
$v(0)$. It should be emphasized that $v(0)$ is not an experimental obse:
able but rather a parameter that was estimated from observables. As
in the case of other estimated quantities, it is necessary to descri
explicitly the methods one uses to obtain estimates of initial, quas:
steady state rates. The fact that many enzymologists omit the descr:
tion of how they obtain initial, quasi-steady state rates is not a pe
in favor of ignoring the description, but rather a compelling reason
question the quality of such results. In the case of individuals whe
know that the methods they use are subjective, but nevertheless do n
describe how they obtain initial rates, one is compelled to question
the motives of the individuals themselves.

The time-honored method of obtaining initial rates is to estima
the initial tangent of $P(t)$ or $S(t)$ visually with a straightedge. Th
method is usually not reproducible from individual to individual (although some individuals claim that is because only they know how to

timate initial tangents properly!) because it includes a step involv-
g human judgement. The method is therefore vulnerable to the bias of
e individual placing the straightedge and even to potential bias intro-
ced by the nature of the straightedge itself. We (Walter and Barrett,
70) have found, for example, that initial slopes estimated visually
om P(t) are generally lower than the estimates obtained from graphical,
merical, or analog computer fits to a suitable mathematical function.
nce these differences are systematically greater in experiments in-
lving lower initial rates even when the reaction order is unchanged,
sual estimates of initial rates can clearly introduce systematic er-
rs.

A better procedure for estimating initial, quasi-steady state rates
to fit P(t) or S(t) directly to a suitable mathematical model, and
lculate the extrapolated initial slope of the function. We (Barrett
d Walter, 1970) have found that for most enzyme kinetic data collected
ring a short time interval of the early stages of the overall reaction,
e MacLaurin polynomial expressing time as a function of product con-
ntration can be truncated to the quadratic form

$$t = c + \eta_0 P + .5\eta_1 P^2 \tag{53}$$

lculation of the three parameters in equation (53) is straightforward:

$$c = (\pi_1 \tau_0 - \pi_2 \tau_1 + \pi_3 \tau_2)/\delta \tag{54}$$

$$\eta_0 = (\pi_5 \tau_1 - \pi_2 \tau_0 - \pi_6 \tau_2)/\delta \tag{55}$$

$$\eta_1 = 2(\pi_3 \tau_6 - \pi_6 \tau_1 - \pi_4 \tau_2)/\delta \tag{56}$$

ere

$$\pi_1 = \Sigma P^2 \Sigma P^4 - (\Sigma P^3)^2$$

$$\pi_2 = \Sigma P \Sigma P^4 - \Sigma P^2 \Sigma P^3$$

$$\pi_3 = \Sigma P \Sigma P^3 - (\Sigma P^2)^2$$

$$\pi_4 = N \Sigma P^2 - (\Sigma P)^2$$

$$\pi_5 = N \Sigma P^4 - (\Sigma P^2)^2$$

$$\pi_6 = N \Sigma P^3 - \Sigma P \Sigma P^2$$

$$\tau_0 = \Sigma t$$

$$\tau_1 = \Sigma t P$$

$$\tau_2 = \Sigma t P^2$$

$$\delta = 2 \Sigma P \Sigma P^2 \Sigma P^3 - N(\Sigma P^3)^2 + N \Sigma P^2 \Sigma P^4 - (\Sigma P^2)^3 - (\Sigma P)^2 \Sigma P^4$$

and N is the total number of data points collected. The variances of these parameters are:

$$VAR(C) = \pi_1 \emptyset / \delta$$

$$VAR(\eta_0) = \pi_5 \emptyset / \delta$$

$$VAR(\eta_1) = \pi_4 \emptyset / \delta$$

here

$$\phi = \frac{\Sigma t \ - c\tau_0 \ -\eta_0\tau_1 \ -\eta_1\tau_2}{N - 3}$$

If we differentiate equation (53) and solve for $\dot{P}$, we obtain

$$\dot{P} = \frac{1}{\eta_0 + \eta_1 P} \tag{60}$$

athematical extrapolation to the initial rate is done by setting P = (0) = 0:

$$v(0) \equiv \dot{P}(0) = \frac{1}{\eta_0} \tag{61}$$

uasi-steady State Data Acquisition

The use of equations (53)-(61) do not require a computer, but the edium of the calculations can be avoided by using some sort of auto-mated calculator. Furthermore, for enzyme reactions wherein a function of P(t) or S(t) can be observed continuously with an instrument (e.g., a spectrophotometer) data acquisition can also be facilitated by using an on-line computer to collect the data. The flow diagram in Figure 8 outlines an automated system (Walter and associates, 1974b; 1975) for the collection and analysis of enzyme kinetic data. In the first step a sample containing all the components necessary for the enzyme-catalyzed reaction but one is placed in the observation path of the detection de-vice. The initial conditions are sampled and stored in the memory of the on-line computer. The reaction is initiated by adding the missing component, and the computer sets the starting time. Data filtration is accomplished by averaging 60 data points per second; the result is stored

Figure 8

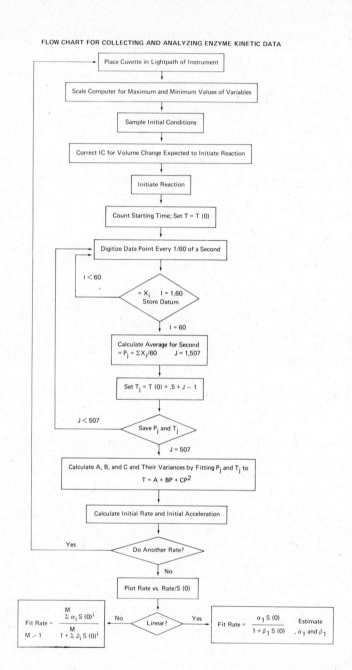

Flowchart for collecting and analyzing enzyme kinetic data.

n the computer memory and interpreted as the datum collected midway

through the second during which the 60 points were collected. When

sufficient averaged data has been collected the reaction is terminated,

and the data is displayed on a digital oscilloscope and stored on a

magnetic tape. The computer eventually uses equations (54)-(61) to fit

the collected data to equation (53), and estimate the variances of the

coefficients and the initial, quasi-steady state rate. Typical results

obtained from a spectrophotometric determination of reduced pyridine

dinucleotide during a dehydrogenase-catalyzed reaction appears in Fig-

ure 9.

Figure 9

```
DATA FROM TAPE UNIT 1.
TYPE 3—DIGIT OCTAL BN WHERE DATA BEGINS.
175
DATA IS FROM THE CARY.
TYPE FULL SCALE OD THEN ZERO SCALE OD:
.5
0
THIS DATA WAS COLLECTED ON 03/03.
THERE ARE +5.080000E+002 DATA POINTS NOT INC THE IC.
HOW MANY POINTS PER CALCULATION?
400
FACTOR TO CONVERT DATA TO MICROMOLAR:
160.8
FIT FOR QUADRATIC:  T = A + BP + CPXP
A = +1.147360E+001 VAR= 3.344730E—001
B = +1.616784E+001 VAR= 6.338327E—002
C = +1.518155E—002 VAR= 2.501916E—003
RATE = +6.185119E—002  INTERVAL = +6.090E—002  TO +6.283E—002
FIT FOR QUADRATIC:  T = A + BP + CPXP
A = +1.179874E+001 VAR= 1.930922E—001
B = +1.607394E+001 VAR= 2.857745E—002
C = +1.938380E—002 VAR= 8.834315E—004
RATE = +6.221251E—002  INTERVAL = +6.157E—002  TO +6.287E—002
   FINISHED WITH THAT DATA.
DO YOU WISH TO ANALYZE ANOTHER SET ON UNIT 1?
Y
OK. LETS GO.
```

Computer output during the collection and analysis of enzyme kinetic
data.

Methods for Distinguishing Enzyme Models

The two simplest mathematical models relating initial, quasi-ste
state rates and one of the initial conditions are

$$v(0) = \frac{\alpha_1 x}{1 + \beta_1 x}$$

(

and

$$v(0) = \frac{\alpha_1 x + \alpha_2 x^2}{1 + \beta_1 x + \beta_2 x^2}$$

(

where x is the initial concentration of a substrate or modifier (i.e.
an "inhibitor" or "stimulator"), and the α and β are constant coefficients that depends on the other initial conditions and the individua
rate constants in the chemical model. More complex models involving
higher order powers of x are clearly possible but equations (62) and
(63) are the two simplest possible models. In what follows we refer
to equation (62) as the "degree-one model" (because it is described b
a rational function of degree one), and to equation (63) as the "degr
two model". For quasi-steady state models the degree-one model resul
from mechanisms such as those involving one binding site on the enzym
(single substrate mechanisms, mechanisms of competitive inhibition,
etc.), multiple, non-interacting binding sites, or ordered sequences
substrate addition. The degree-two model results from mechanisms involving interacting binding sites on the enzyme (not competitive (Wal
1962) inhibition, not competitive stimulation, random or preferred or
mechanisms, etc.).

The minimum aim of any scientific investigation should be to distinguish between the two simplest mathematical models. In the case o

uasi-steady state kinetic studies with enzyme-catalyzed reactions, the
oal is to decide if the data fit equation (62), or if equation (63)
or a rational function of higher degree) must be used.   The decision
s easy to make if the experimental data contain a feature such as an
xtremum or an inflection: Mechanisms described by equation (62) can-
ot have an extremum or an inflection in plots of $v(0)$ versus x, but
echanisms described by equation (63) have an extremum if

$$\frac{\partial v(0)}{\partial x} = \frac{\alpha_1 + 2\alpha_2 x + (\alpha_2\beta_1 - \alpha_1\beta_2)x^2}{(1 + \beta_1 x + \beta_2 x^2)^2} \tag{64}$$

or positive, real x.   In other words if $\alpha_1\beta_2 > \alpha_2\beta_1$, the degree-two
odel possesses a feature (an extremum) that cannot be present in the
egree-one model.   Furthermore if

$$\frac{\partial^2 v(0)}{\partial x^2} = \frac{\alpha_2 - \alpha_1\beta_1 - 3\alpha_1\beta_2 x - 3\alpha_2\beta_2 x^2 - \beta_2(\alpha_2\beta_1 - \alpha_1\beta_2)x^3}{(1 + \beta_1 x + \beta_2 x^2)^3} \tag{65}$$

the degree-two model possesses an inflection point in plots of $v(0)$ ver-
sus x.   If $\alpha_1\beta_2 > \alpha_2\beta_1$ there must be an inflection between the extremum
and asymptotic values of $v(0)$.   If $\alpha_1\beta_2 > \alpha_2\beta_1$ and $\alpha_2 > \alpha_1\beta_1$, there is
also an inflection between zero and the extremum value of $v(0)$.   Finally
if $\alpha_1\beta_2 < \alpha_2\beta_1$, there is no extremum, but if $\alpha_2 > \alpha_1\beta_1$, there is an inflection point.   In all these cases the degree-two model possesses a
feature that distinguishes it from the degree-one model.   Therefore if
experimental data indicate that plots of $v(0)$ versus x possess an extremum or an inflection, the degree-one model can be eliminated from
consideration.
     It is clear from equations (62) and (63) that the degree-two model
converges to the degree-one model as α_2 and β_2 approach zero.   This

means that for small α_2 and β_2 it will be difficult or impossible to distinguish between these two models. However, in such cases the numerical values for α_1 and β_1 calculated for each model will be similar.

Even if α_2 and β_2 are relatively large, it may not always be possible to distinguish between a hyperbola generated from equation (63) when $\alpha_1\beta_2 < \alpha_2\beta_1$ and $\alpha_2 < \alpha_1\beta_1$ and a hyperbola generated for a different set of α_1 and β_1 from equation (62). Statistical fits of real data to equations (62) and (63) may lead to different sets of parameters, each of which possesses a minimum sum of squared deviations. When the minimum sums produced by the fit to each model are sufficiently similar that they cannot be experimentally distinguished, the two models become indistinguishable. In this case however, the numerical values of α_1 and β_1 calculated for each model will be different.

The usual method of distinguishing between the degree-one model and higher degree models like equation (63) is to plot the experimental quantities in a "biased" form that should be linear from the degree-one model, but nonlinear for the degree-two model. In such cases it should be possible to detect a specific concave trend if the biased plot is not linear. The use of graphical procedures of this type depend upon finding a form that is linear for one model, but markedly nonlinear for the other model.

Equation (62) can be rearranged to the linear forms:

$$\frac{1}{v(0)} = \frac{\beta_1}{\alpha_1} + \frac{1}{\alpha_1}\frac{1}{x} \tag{6}$$

$$\frac{x}{v(0)} = \frac{1}{\alpha_1} + \frac{\beta_1}{\alpha_1}x \tag{6}$$

$$v(0) = \frac{\alpha_1}{\beta_1} - \frac{1}{\beta_1}\frac{v(0)}{x} \tag{6}$$

quation (66) illustrates that for the degree-one model, plots of $1/v(0)$
ersus $1/x$ should be linear; the $1/v(0)$ axis intercept is β_1/α_1, and
he slope is $1/\alpha_1$. Equation (67) illustrates that for this model plots
f $x/v(0)$ versus x should also be linear; the $x/v(0)$ axis intercept is
$/\alpha_1$, and the slope is β_1/α_1. And equation (68) illustrates that plots
f $v(0)$ versus $v(0)/x$ for the degree-one model should be linear; the
(0) axis intercept is α_1/β_1, and the slope is $-1/\beta_1$.

If one attempts to rearrange equation (63) into any of the forms
that are linear for equation (62), one does not obtain linear relationships. Hofstee (1952) pointed out that the reciprocal method of plot-
ting initial rates and substrate concentrations could obscure deviations
from equation (62) and he emphasized the advantages of $v(0)$ versus
$v(0)/x$ plots to detect model errors. Notwithstanding these criticisms
(Walter, 1965), 97.3 percent of the "linear plots" in the 1971 issue
of the _Journal of Biological Chemistry_ were reciprocal plots; only 1.6
percent were the demonstrably superior $v(0)$ versus $v(0)/x$ plots (Walter,
1974c), and the remaining 1.1 percent were the grossly inferior $x/v(0)$
versus x type plot. Since 20.2 percent of all papers in the 1971 issue
of the _Journal of Biological Chemistry_ uses one of these linear forms
and since each of these papers contains an average of 10.1 linear plots,
the relationship between initial, quasi-steady state rates and initial
substrate concentrations must be vital to the results published in this
prestigeous journal. This preference by enzymologists for reciprocal
plots (even though such plots might obscure information in their results) reminds one of the old adage about how difficult it is to teach
old dogs new tricks.

$v(0)$ versus $v(0)/x$ plots are superior for purposes of distinguishing
equations (62) and (63) because the curvature in $v(0)$ versus $v(0)/x$
plots of initial rates from the degree-two model is greater than the
curvature in reciprocal plots or in plots of $x/v(0)$ versus x for the
same model. For the degree-two model the curvatures are

$$\frac{\partial^2 \frac{1}{v(0)}}{\partial \left(\frac{1}{x}\right)^2} = \frac{2x^3[\alpha_2^2 + \alpha_1(\alpha_1\beta_2 - \alpha_2\beta_1)]}{(\alpha_1 + \alpha_2 x)^3},$$

(

$$\frac{\partial^2 \left(\frac{x}{v(0)}\right)}{\partial x^2} = \frac{2[\alpha_2^2 + \alpha_1(\alpha_1\beta_2 - \alpha_2\beta_1)]}{(\alpha_1 + \alpha_2 x)^3},$$

(

and

$$\frac{\partial^2 v(0)}{\partial \left(\frac{v(0)}{x}\right)^2} = -\frac{2[\alpha_2^2 + \alpha_1(\alpha_1\beta_2 - \alpha_2\beta_1)](1 + \beta_1 x + \beta_2 x^2)^3}{(\alpha_1\beta_1 - 2\alpha_1\beta_2 x + \alpha_2\beta_2 x^2)^3}.$$

(

In terms of the scaled quantities,

$$\sigma \equiv \alpha_1 x \quad \text{and} \quad V \equiv v(0)\beta_2/\alpha_2$$

the curvature equations are

$$c_1 \equiv \frac{\partial^2 \left(\frac{1}{V}\right)}{\partial \left(\frac{1}{\sigma}\right)^2} = \frac{2a_2\sigma^3[a_2 + a_1(a_1 - 1)]}{(a_1 + a_2\sigma)^3},$$

(7

$$c_2 \equiv \frac{\partial^2 \left(\frac{\sigma}{V}\right)}{\partial \sigma^2} = \frac{2a_2[a_2 + a_1(a_1 - 1)]}{(a_1 + a_2\sigma)^3}$$

(7

and

$$c_3 \equiv \frac{\partial^2 V}{\partial \left(\frac{V}{\sigma}\right)^2} = -\frac{2a_2[a_2 + a_1(a_1 - 1)](1 + \sigma + a_2\sigma^2)^3}{(a_1 - a_2 + 2a_1a_2\sigma + a_2^2\sigma^2)^3}$$

(7

$$a_1 = \frac{\alpha_1 \beta_2}{\alpha_2 \beta_1} < 1 \quad \text{and} \quad a_2 = \frac{\beta_2}{\beta_1^2} < a_1$$

The absolute value of the ratio of the curvature in plots of $1/V$ versus $1/\sigma$ to the curvature in plots of V versus V/σ is

$$\frac{C_1}{C_3} = \frac{\sigma^3(a_1 - a_2 + 2a_1 a_2 \sigma + a_2^2 \sigma^2)^3}{(1 + \sigma + a_2 \sigma^2)^3(a_1 + a_2 \sigma)^3} \tag{75}$$

aking the cube root and multiplying out the terms, we find

$$\sqrt[3]{\frac{C_1}{C_3}} = \frac{(a_1 - a_2)\sigma + 2a_1 a_2 \sigma^2 + a_2^2 \sigma^3}{a_1 + (a_1 + a_2)\sigma + a_2(a_1 + 1)\sigma^2 + a_2 \sigma^3} < 1 \tag{76}$$

here $a_1 + a_2 > a_1 - a_2$ because $a_2 > 0$, and $a_1 + 1 > 2a_1$ because $a_1 < 1$. nequality (76) illustrates that for all $0 \leq \sigma < \infty$ in the degree-two odel, plots of V versus V/σ possess more curvature than do plots of $/V$ versus $1/\sigma$ constructed from the same data (Walter, 1974c).

The absolute value of the ratio of the curvature in plots of σ/V ersus σ to the curvature in plots of V versus V/σ is

$$\frac{C_2}{C_3} = \frac{(a_1 - a_2 + 2a_1 a_2 \sigma + a_2^2 \sigma^2)^3}{(1 + \sigma + a_2 \sigma^2)^3(a_1 + a_2 \sigma)^3} \tag{77}$$

n this case we find that

$$\sqrt[3]{\frac{C_2}{C_3}} = \frac{a_1 - a_2 + 2a_1 a_2 \sigma + a_2^2 \sigma^2}{a_1 + (a_1 + a_2)\sigma + a_2(a_1 + 1)\sigma^2 + a_2^2 \sigma^3} < 1 \tag{78}$$

where $a_1 > a_1 - a_2$ because $a_2 > 0$, and $a_1 + a_2 > 2a_1a_2$ and $a_1 + 1 > a$

because $1 > a_1 > a_2$. Inequality (78) illustrates that for all $0 > \sigma$

$> \infty$ in the degree-two model, plots of V versus V/σ possess more curv

ature than do plots of σ/V versus σ constructed from the same data

(Walter, 1974c).

Table 1 contains a list of the minimum and maximum possible slop

<u>TABLE 1</u>

TYPE PLOT	SLOPE		CURVATURE	
	Low σ	High σ	Low σ	High σ
$\frac{1}{V}$ vs $\frac{1}{\sigma}$	$\frac{1}{a_1}$	$\frac{1 - a_1}{a_2}$	$\longrightarrow 0$ minimum	$\frac{2(a_2 + a_1(a_1 - 1))}{a_2^2}$ maximum
$\frac{\sigma}{V}$ vs σ	$\frac{a_1 - a_2}{a_1}$	1	$\frac{2a_2(a_2 + a_1(a_1 - 1))}{a_1^3}$ maximum	$\longrightarrow 0$ minimum
V vs $\frac{V}{\sigma}$	$\frac{a_1}{a_2 - a_1}$	$\frac{a_1 - 1}{a_2}$	$\frac{2a_2(a_2 + a_1(a_1 - 1))}{(a_2 - a_1)^3}$ minimum	$\frac{-2(a_2 + a_1(a_1 - 1))}{a_2^2}$ maximum

$\sigma = a_1 x$ $V = v(0)\beta_2/\alpha_2$ $a_1 = \alpha_1\beta_2/\alpha_2\beta_1$ $a_2 = \beta_2/\beta_1^2$

Maximum curvatures and slopes for transformed plots of the variances

equation (63).

and curvatures for the degree-two model when V and σ are plotted in

ıch of the forms that would be linear for the degree-one model. In-
ıualities (76) and (78) and Table 1 illustrate that the V versus V/ɑ
ɔr the v(0) versus v(0)/x) type plot is the best type to use to detect
ɔviations from the degree-one model.

Figure 10 illustrates the case when it is not possible to distin-
ıish between hyperbolas generated from equations (62) and (63). The

Figure 10

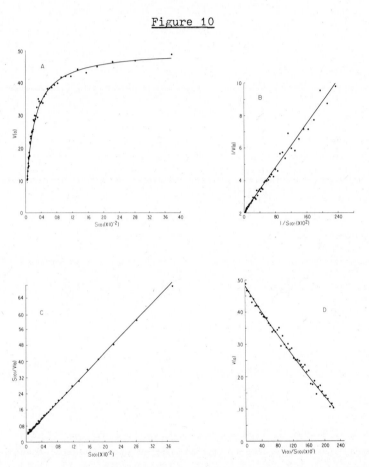

The values of v(0) were generated with random error (ten percent
ʰhen v(0) equals one-tenth its asymptotic value, one percent when v(0)
ɋquals its asymptote, and proportionate in between) from equation (63)
ʰhen $\alpha_1 = 3 \times 10^3$, $\beta_1 = 10^4$, $\alpha_2 = 10^7$, and $\beta_2 = 2 \times 10^7$. The solid lines
ɾe the exact function (equation (63)) with no error.

actual model is equation (63), $\alpha_1 = 3 \times 10^3$, $\beta_1 = 10^4$, $\alpha_2 = 10^7$, and β

2×10^7. In Figure 10A the initial rates generated with typical expe

imental error for this model are plotted versus x, the initial substr

concentration. The solid line in this figure is the plot of v(0) ver

x generated directly from equation (63) (without error). Note that t

data look very much like what one would expect for the degree-one mod

Figures 10B, 10C, and 10D are the three types of plots that shou

be linear for equation (62). The plotted quantities are the transfor

variables calculated from the plotted quantities in Figure 10A, and t

solid lines are the values of the transformed variables generated dir

ly from equation (63). Note that all three plots appear linear: the

data still look very much like what one would expect for the degree-o

model. However, the value of $\beta_1 = 6.2 \times 10^3$ calculated for the degre

one model from a weighted least squares fit to the linear forms of eq

tion (62) is different from the value used to generate the data from

equation (63). Since the significance level* for the distinction be-

tween equations (62) and (63) calculated from the variance ratio for

two models and fifty experimental determinations is in the range of f

percent, these two models cannot be distinguished when $\alpha_1 = 3 \times 10^3$ a

$\beta_1 = 6.2 \times 10^3$ for the degree-one model, and $\alpha_1 = 3 \times 10^3$, $\beta_1 = 10^4$,

$\alpha_2 = 10^7$, and $\beta_2 = 2 \times 10^7$ for the degree-two model.

Figure 11 illustrates the case when it is relatively easy to dis

tinguish between hyperbolas generated from equations (62) and (63).

actual model is again equation (63), but in Figure 11 $\alpha_1 = 2 \times 10^3$,

$\beta_1 = 8 \times 10^3$, $\alpha_2 = 2 \times 10^6$, and $\beta_2 = 4 \times 10^6$. The initial rates gene

ated with typical experimental error for this model are plotted versu

x in Figure 11A. The solid line is the plot of v(0) versus x generat

directly from equation (63) without error. Note that the data in Fig

---

*The significance level is the number of times out of a hundred that
data will lead to the correct model. If one flips a coin to decide b
tween equations (62) and (63), the significance level of the decision
would be 50 percent.

Figure 11

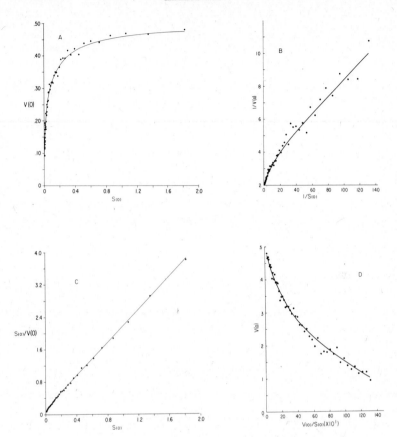

The values of $v(0)$ were generated as in Figure 10 except $\alpha_1 =$ x 10^3, $\beta_1 = 8$ x 10^3, $\alpha_2 = 2$ x 10^6, and $\beta_2 = 4$ x 10^6. The solid lines .re the exact function with no error.

1A look very much like what one would expect for the degree-one model.

Figures 11B, 11C, and 11D are the three type plots that should be .inear for equation (62). The plotted quantities are the transformed variables calculated from the plotted quantities in Figure 11A, and the solid lines are values of the transformed variables generated directly from equation (63). In this case the $1/v(0)$ versus $1/x$ plot (Figure 11B) .s curved, but the curvature is not obvious at lower $v(0)$. The $x/v(0)$ versus x plot (Figure 11C is also curved, but the curvature is not

obvious at all. The v(0) versus v(0)/x plot (Figure 11D) possesses
the most obvious curvature of the three types of transformed plots,
and it is therefore the best type to use to detect deviations from eq
tion (62). The significance level for the distinction between equati
(62) and (63) based on fifty experimental determinations is 99.9, and
the significance level based on twelve experimental determinations of
v(0) is 91%. Thus, when $\alpha_1 = 2 \times 10^3$, $\beta_1 = 8 \times 10^3$, $\alpha_2 = 2 \times 10^6$,
$\beta_2 = 4 \times 10^6$, one can use v(0) versus v(0)/x plots or statistical cal
culations of significance levels to ascertain that the data (Figure 1
fit equation (63) much better than they fit equation (62), but the st
tistical calculations depend on the availability of what in practical
terms is a rather large number of experimental determinations.

Figure 12 illustrates a case when it is relatively difficult to
distinguish between hyperbolas generated from equations (62) and (63)
The actual model is once again equation (63), but in Figure 12 $\alpha_1 =$
2×10^3, $\beta_1 = 8 \times 10^3$, $\alpha_2 = 2 \times 10^6$, and $\beta_2 = 5.5 \times 10^6$. The initial
rates generated with typical experimental error for this model are pl
ted versus x in Figure 12A. The solid line is the plot of v(0) versu
x generated directly from equation (63) without error.

Figures 12B, 12C, and 12D are again the three type plots that sh
be linear for equation (62). The solid lines are the unweighted regr
sion fits to the respective linear forms of equation (62). In this c
only the v(0) versus x(0)/x plot (Figure 12D) possesses obvious curvat
The average significance level for distinguishing the two models on t
basis to ten to a hundred determinations appears in Figure 13. These
significance levels range from 70 percent at ten data points up to 95
percent at a hundred points. This means that to obtain the usual lev
of significance to distinguish models on the basis of statistical cal
culations, one would have to determine over a hundred initial rates.

In Figure 14 appears the relationship between σ and the absolute
value of the normalized curvature for the solid lines in Figures 10B

Figure 12

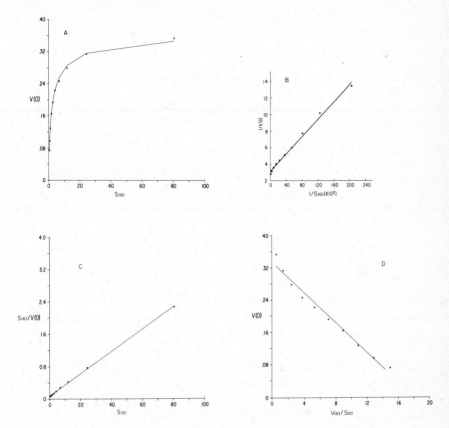

The values of $v(0)$ were generated as in Figure 11 except $\beta_2 =$ 5 x 10^6. The solid line in 12A is the exact function with no error, nd the solid lines in 12B - 12D are the unweighted regression fits to he linear forms of equation (62).

l/V versus 1/σ), 10C (σ/V versus σ) and 10D (V versus V/σ). As antic-pated from inequalities (76) and (78), the curvature in plots of V ver-/σ is always larger than the curvature in plots of 1/V versus 1/σ or /V versus σ constructed from the same data.

It is clear from these illustrations that the nonlinear trends riginating from the degree-two model are most obvious in the $v(0)$ ver-is $v(0)/x$ type plot. Furthermore, $v(0)$ versus $v(0)/x$ plots can be more

Figure 13

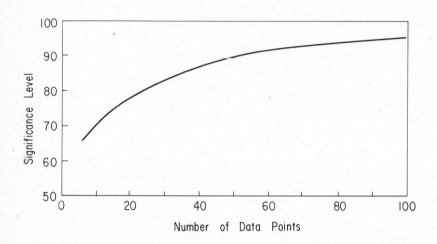

The average significance levels were calculated from twenty-five sets of v(0) and S(0) for each number of data points. The individual significance levels were calculated from the ratio of the variances ob tained by fitting the points to equations (62) and (63). Each set of v(0) and S(0) used to construct this figure was generated independentl as described in Figure 12.

useful for distinguishing equations (62) and (63) than statistical cal culations of significance levels.

Figure 14

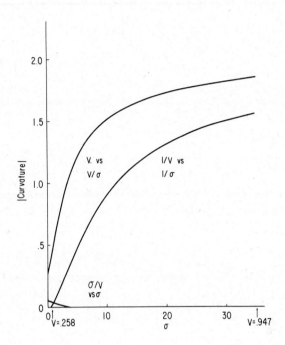

Plot of the absolute value of the curvatures calculated from equations (72) - (74) versus $\alpha_1 S(0)$ for $\alpha_1 = 3 \times 10^3$, $\beta_1 = 10^4$, $\alpha_2 = 10^7$, and $\beta_2 = 2 \times 10^7$.

VI. Cooperativity in Enzyme-catalyzed Reactions

The importance of distinguishing between enzymes that follow the
hyperbolic rate law described by equation (62) and those that do not
is underscored by the important role that some enzymes in the latter
category play in the control of metabolic processes. It is often fou
for example, that one or more key enzymes in a biochemical pathway ca
under the proper conditions, exhibit rate laws wherein the relationsh
between the initial rate and the initial substrate concentration involves an inflection and/or an extremum. Often such nonhyperbolic en
zymes play a key role in the overall control of their metabolic pathw

A sound theoretical basis for nonhyperbolic rate laws for these
enzymes was developed nearly twenty years ago by Jean Botts and assoc
ates (1957; 1958; 1959). In this formulation a sigmoidal relationshi
between the initial rate and the initial substrate concentration was
shown to be one of several possible relationships for enzymes involvi
ternary complexes. Botts (1958) pointed out that enzymes forming hig
order compounds could also result in sigmodal or other nonhyperbolic
rate laws. Subsequently, Monod and associates (1965) suggested the "a
losteric" model as a theoretical basis for the sigmoidal relationship
between oxygen binding to hemoglobin and oxygen concentration. In the
general model the protein is composed of $n/2 = \rho$ subunits, each of whi
possesses one binding site for the ligand. Furthermore, each alloster
protein exists in two forms, R and T, such that the equilibrium consta
for the reaction

$$T \rightleftarrows R$$

is L. The equilibrium constant for the binding of the first molecule
of ligand to the R form is ρK_r; the corresponding constant for the bin
ing to the T form is ρK_t. We now define $\alpha = K_r s$ and $\beta = K_t/K_r$, where

s the concentration of the ligand. At equilibrium the fraction of the
inding sites on the protein occupied by a molecule of ligand is

$$F = \frac{\alpha\beta(1 + \alpha\beta)^{\rho-1} + L\alpha(1 + \alpha)^{\rho-1}}{L(1 + \alpha)^{\rho} + (1 + \alpha\beta)^{\rho}}. \tag{79}$$

The model postulated for hemoglobin is a nonkinetic model wherein
ll the chemical components are assumed to be in thermodynamic equili-
rium with one another. Specifically, hemoglobin is not an enzyme, and
nlike enzyme-substrate compounds, the hemoglobin compounds do not "turn-
ver" to form a product from the oxygen. There is only binding, no turn-
ver (Walter, 1970; 1972; 1973). Therefore extrapolation of the results
btained from the theoretical hemoglobin model (equation (79)) to enymes should be limited to situations wherein quasi-equilibrium (See
ection I of this chapter) is strictly maintained. In the general enyme model in Figure 15, the enzyme is composed of $n/2 = \rho$ subunits,
ach of which possesses one binding site for the substrate. The T forms
f the enzyme are denoted by X's with odd subscripts, and the R forms
y X's with even subscripts. In this model strict quasi-equilibrium
eans that

$$l_i \gg h_i \ll k_{-i} \qquad i = 1,\ldots,n \tag{80}$$

pplication of the "allosteric" model (equation (79)) to enzyme-catalyzed
eactions (Figure 15) is implicitly limited to the conditions specified
n inequalities (80) (Walter, 1970; 1972). In what follows we shall
se v_e to denote all such initial rates calculated on the basis of the
uasi-equilibrium assumption.

When quasi-equilibrium is not maintained (the usual situation with
nzymes), it is necessary to solve the quasi-steady state equations in
rder to ascertain the relationship between the initial, quasi-steady

Figure 15

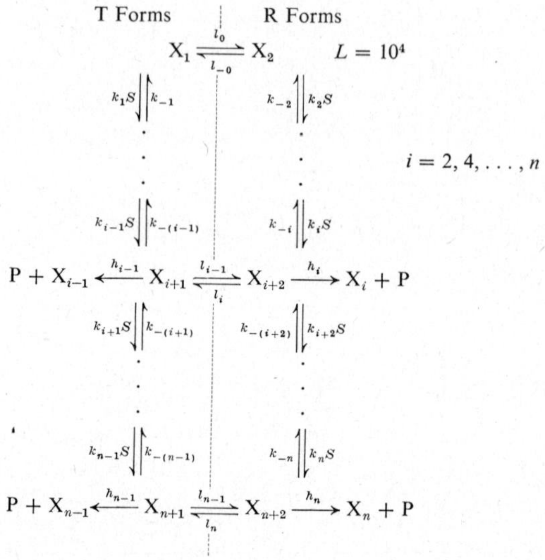

$$
\begin{array}{ccc}
\text{T Forms} & \vdots & \text{R Forms} \\
& X_1 \underset{l_{-0}}{\overset{l_0}{\rightleftharpoons}} X_2 & L = 10^4 \\
k_1 S \updownarrow k_{-1} & & k_{-2} \updownarrow k_2 S \\
\vdots & & \vdots \\
& & i = 2, 4, \ldots, n \\
k_{i-1}S \updownarrow k_{-(i-1)} & & k_{-i} \updownarrow k_i S \\
P + X_{i-1} \xleftarrow{h_{i-1}} X_{i+1} \underset{l_i}{\overset{l_{i-1}}{\rightleftharpoons}} X_{i+2} \xrightarrow{h_i} X_i + P \\
k_{i+1}S \updownarrow k_{-(i+1)} & & k_{-(i+2)} \updownarrow k_{i+2}S \\
\vdots & & \vdots \\
k_{n-1}S \updownarrow k_{-(n-1)} & & k_{-n} \updownarrow k_n S \\
P + X_{n-1} \xleftarrow{h_{n-1}} X_{n+1} \underset{l_n}{\overset{l_{n-1}}{\rightleftharpoons}} X_{n+2} \xrightarrow{h_n} X_n + P
\end{array}
$$

Chemical model for an enzyme with P subunits.

state rate, v_{ss}, and the initial substrate concentration. This has be
done for the enzyme model in Figure 15 elsewhere (Walter, 1972); here
we shall summarize some of the more interesting results.

Since $K_m \geq K_s$, the maximum value possible for v_{ss} is v_e obtained
at the same $S(0)$. Therefore, if all the k's in Figure 15 remain unchanged in the comparison, quasi-equilibrium rates are limiting values
of quasi-steady state rates, and sigmoidal relationships between v_e an
$S(0)$ are limiting cases of the corresponding sigmoidal relationships
between v_{ss} and $S(0)$. This is illustrated in Figure 16: In Figure 16

Figure 16

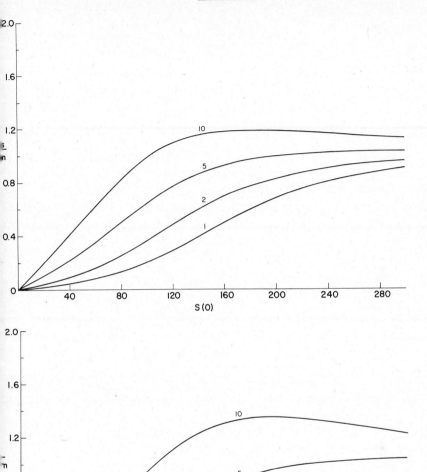

Relationship between the initial, quasi-steady state rate (Top
urves) or the corresponding initial, quasi-equilibrium rate (Bottom
urves) of an enzyme-catalyzed reaction and the initial substrate con-
entration. The numbers above each curve are the ratios of h_i (i odd)
o h_j (j even) in Figure 15.

appear several sigmoidal relationships between v_{ss} and $S(0)$, each obtained at a different ratio of h_i/h_j; Figure 16b illustrates the corresponding limiting sigmoidal relationships obtained at quasi-equilibrium.

Figure 16 also illustrates that, whether or not quasi-equilibriu
is maintained, an enzyme displaying a sigmoidal relationship between
the initial rate and $S(0)$ in the absence of a modifier could be made
display a much less sigmoidal and very nearly hyperbolic relationship
between these same variables when a modifier that inhibits the h_j (j
even), but not the h_i (i odd) is added. Similarly, an enzyme display
a nearly hyperbolic relationship in the absence of a modifier could b
made to display a sigmoidal relationship when a modifier that stimula
the h_j (j even), but not the h_i (i odd) is added.

Figure 17 illustrates the effect of increasing remoteness from
quasi-equilibrium on the sigmoidal relationship between initial rates
and initial substrate concentrations. In the five curves all the rat
constants except the h_i(i = 1,2,...,n) remain unchanged. In the cur
labeled 10^{-6}, $h_i/k_{-i} = 10^{-6}$(i = 1,2,...,n), inequalities (80) are sat
fied, and quasi-equilibrium is maintained. In the remaining curves
h_i/k_{-i}(i = 1,2,...,n) is the value indicated for each curve (0.5, 1.0
2.0, or 5.0), inequalities (80) are not satisfied, and quasi-equilibrium is not maintained. Thus the top curve in Figure 17 is the limi
ing case for the family of quasi-steady state curves appearing under

Figure 18 illustrates the effect of the ratios of h_i/h_j (i odd,
even) on the relationship between the initial rate and the initial su
strate concentration when quasi-equilibrium is not maintained, and it
is the h_j with the even subscripts (L = 10^4) that are not changed. I
Figure 18a the h_j (j even) are twice the h_j (j even) in Figure 18b.
the curves labeled 1, the h_i (i odd) equal the h_j (j even); in the cu
labeled 2, the h_i (i odd) are twice the h_j (j even); in curve 5 the h
(i odd) are five times the h_j (j even); in curve 10 the h_i (i odd) ar

Figure 17

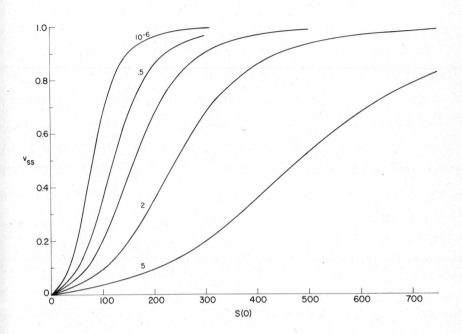

Relationship between the initial rate of an enzyme-catalyzed reaction and the initial substrate concentration. The numbers above each
curve are h_i/k_{-i}, i = 1,2,...,n, an indicator of how remote the enzyme
mechanism in Figure 15 is from quasi-equilibrium.

en times the h_j (j even); in each curve the h_i with odd subscripts are
equal; in the top set of curves in Figure 18a the h_j with even subscripts
re all equal to each other and are twice the values of the h_j with even
ubscripts in the bottom set of curves. Figure 18 illustrates that an
nzyme displaying a sigmoidal relationship between v_{ss} and S(0) in the
bsence of a modifier could be made to display a much less sigmoidal and
ery nearly hyperbolic relationship between these same variables when a

Figure 18

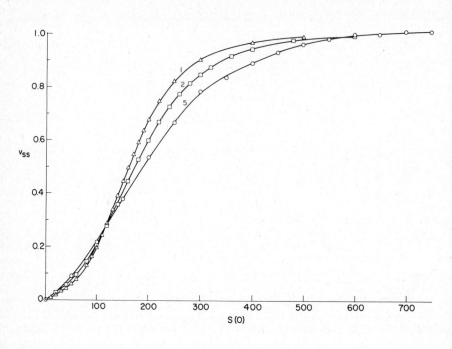

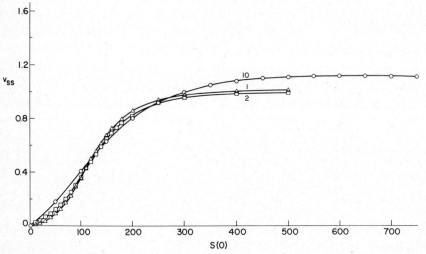

Relationship between the initial, quasi-steady state rate of an enzyme-catalyzed reaction and the initial substrate concentration. T numbers above each curve are the ratios of h_i (i odd) to h_j (j even) Figure 15.

difier that stimulates the h_i (i odd), but not the h_j (j even) is
.ded. Similarly, an enzyme displaying a "normal" hyperbolic relation-
.ip could be made to exhibit a sigmoidal relationship between v_{ss} and
0) when a modifier that inhibits the h_i (i odd), but not the h_j (j
·en) is added. These results, together with those illustrated in Fig-
·e 16, are summarized in Table 2.

TABLE 2

Summary of Results Illustrated in Figures 16 and 18

difier* Effect on te Constant for rnover of Enzyme		Possible Effect on Initial Rate Versus Initial Substrate Concentration Curve
(i odd)	h_j (j even)	
ne	Inhibit	Sigmoidal $\rightarrow$ hyperbolic
timulate	None	Sigmoidal $\rightarrow$ hyperbolic
ne	Stimulate	Hyperbolic $\rightarrow$ sigmoidal
hibit	None	Hyperbolic $\rightarrow$ sigmoidal

s Professor Manuel Morales pointed out at the 1970 Biophysical Society
meetings, the effect on the rate constant need not be due to a chemical
:omponent. It is well known, for example, that a change in temperature
:an result in a decrease or an increase in the value of a rate constant.

Figure 19 illustrates the effect of altering the individual k_j and
$_{-j}$ (j even) in such a manner that the binding constant, $K_j = k_j/k_{-j}$,
s itself unchanged. Thus all the constants in the three curves in Fig-
·e 19 are the same except k_j and k_{-j} (j even), but their ratio, K_j is
lso the same in each curve. In curve 1, h_j/k_{-j} (j even) is 1; in curve
), h_j/k_{-j} (j even) is 10, but h_j is the same as in curve 1 and k_j is
s one tenth of the value used in curve 1; in curve 100, h_j/k_{-j} (j even)
s 100, and k_j is one hundredth of the value used in curve 1. Thus the
ffect illustrated in Figure 19 is strictly on binding and therefore is

Figure 19

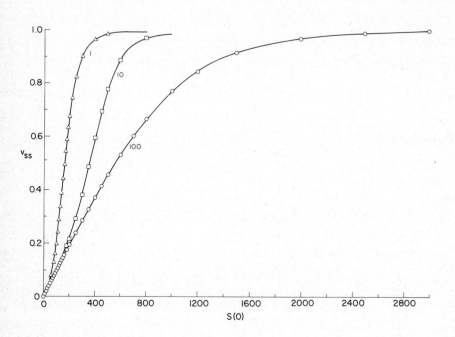

The effect of the individual k_j and k_{-j} (j even) on the relation
ship between the initial, quasi-steady state rate of a reaction cata
yzed by the enzyme described in Figure 15 and its initial substrate
concentration. The numbers above each curve are the ratios of h_j to
k_{-j} (j even) in Figure 15, but in all curves $K_j = k_j/k_{-j}$ is constant.

due to the fact that quasi-equilibrium is not maintained. This purel
kinetic effect is to shift sigmoidal curve 1 (which is 50% saturated
when S(0) is about 160) to the much less sigmoidal and very nearly hy
bolic curve 100 (which is 50% saturated when S(0) is about 560). The
fect in Figure 19 is particularly remarkable in view of the fact that
mechanism, all the binding constants, and all the rate constants for

nover of enzyme-substrate compounds are identical in the three curves.
s result illustrates that a substance that effects neither the bind-
constants for an allosteric enzyme nor the rate constants for the
nover of the enzyme-substrate compounds can nevertheless cause a
ft from a sigmoidal relationship between initial rates and initial
strate concentrations to a rectangular hyperbolic relationship or
e versa.

Bibliography

erty, R. A. and Miller, W. G. (1958). J. Am. Chem. Soc. 80 5146.
rett, M. J. and Walter, C. F. (1970). Enzymol. 38 161.
enstein, M. (1913). Z. Physik. Chem. 85 329.
ts, J. and Drain, C. (1957). Conference on the Chemistry of Muscular
traction, Igaku Shoin Ltd Publishers, Tokyo.
ts, J. (1958). Trans. Faraday Soc. 54 593.
ggs, G. E. and Haldane, J. B. S. (1925). Biochem. J. 19 383.
wn, A. (1902). J. Chem. Soc. 81 373.
her, D. D. and Schultz, A. R. (1969). Math. Biosciences 4 189.
ron, J., Bernhard, S., Friess, S., Botts, D. and Morales, M. F. (1959)
"The Enzymes" (P. Boyer, H. Lardy and M. Myrbäck, eds). Academic
ss, New York, Chapter 2.
neken, F. G., Tsuchiya, F. G. and Aris, R. (1967). Math. Biosciences
5.
ri, V. (1903) in "Lois générales de l'action des diastases", Hermann
ss, Paris
stee, B. H. J. (1952). Science 116 329.
rson, K. E. (1962) in "A Programming Language", Wiley Press, New York.
rson, K. E. (1963). Comm. Assoc. Comp. Mach. 7 80.
rson, K. E. (1964). IBM Systems J. 2 117.
g, L. and Altman, C. (1956). J. Phys. Chem. 60 1375.
ne, W. (1878) Untersuch. Physiol. Inst. Univ. Heidelberg 1 291.
haelis, L. and Menten, M. (1913). Biochem Z. 49 333.
od, J., Wyman, J., and Changeux, J. (1965). J. Mol. Biol. 12 88.
ales, M. F. and Goldman, D. (1955). J. Am. Chem. Soc. 77 6069.
ultz, A. R. and Fisher, D. D. (1969). Can. J. Biol. Chem. 47 889.
honov, A. N. (1952). Mat. Sb. 31 (73) 575.
ter, C. F. (1962). Biochemistry 1 652.
ter, C. F. and Morales, M. F. (1964). J. Biol. Chem. 239 1277.
ter, C. F. (1965) in "Steady-state Applications in Enzyme Kinetics",
ald Press, New York.
ter, C. F. (1966). J. Theor. Biol. 15 1.
ter, C. F. (1970). Proc. Biophys. Soc. 14 120a.
ter, C. F. and Barrett, M. J. (1970). Enzymol. 38 147.
ter, C. F. (1972) in "Biochemical Regulatory Mechanisms in Eukaryotic
lls" (E. Kun and S. Grisola, eds) Wiley Interscience Publishers, New
k, Chapter 11.
ter, C. F. (1973). Symp. Biogenesis-Evolution-Homeostasis pg 51.
ter, C. F. (1974a). J. Theor. Biol. 44 1.
ter, C. F., Eberspaecher, H. and Hughes, J. P. (1974b). Int. J.
antum Chem.: Quant. Biol. Symp. 1 253.

Walter, C. F. (1974c). J. Biol. Chem. 249 699.
Walter, C. F., Eberspaecher, H. and Hughes, J. P. (1975). Anal. Bi
69 590.
Wong, J. T. (1965). J. Am. Chem. Soc. 87 1788.
Yang, C. (1954). Arch. Biochem. Biophys. 51 419.

In the next paper the biological level of organization shifts from
e molecular to the cellular level. Professor Kauffman briefly describes
o old models of the mitotic cycle, and illustrates how they have con-
.buted to the development of a more adequate theory. The new model,
.ch is based on the nonlinear properties of sustained biochemical os-
.lations, has been used ingeniously by Professor Kauffman and his as-
:iates to suggest experimental tests of the theory. The contributions
the nonlinear dynamic model and these experimental tests of the model
our understanding of the control of mitosis are described in the fol-
ving paper.

DYNAMIC MODELS OF THE MITOTIC CYCLE:
EVIDENCE FOR A LIMIT CYCLE OSCILLATOR

Stuart Kauffman M. D.
Associate Professor
Dept. of Biochemistry and Biophysics
School of Medicine
University of Pennsylvania
Philadelphia, PA

I. Introduction: Models of the Mitotic Cycle

Control of the periodicity of mitosis, and maintenance of stabl
phase relation among events of the cell cycle, have been extensively
studied in a number of prokaryotes (1,2) in tetrahymena (3,4) Parame
(5), yeast (6), Physarum (7,8) and higher cells (9). Two broad clas
of models have emerged over the past several decades (reviewed by J
Mitchison 10). The first envisions a recurrent sequence of discrete
cellular events, related to one another as a simple causal loop sequ
or as partially independent, partially connected causal sequences ev
ually forming a loop. Hartwell's (11) elegant genetic analysis of t
yeast division cycle using temperature sensitive mutants stopping ce
at specific phases has led him to this type of model. Such a causal
loop model would explain both the <u>periodicity</u> of the cycle and the m
tenance of proper phase relations between events of the cycle. The
of the cycle are here considered as parts of the clock.

The second broad class of models proposes the continuous accumu
of a mitogen, or division protein during the cell cycle which reache
critical concentration, perhaps converts to a new division structure
initiates mitosis, is used up in mitosis, and reaccumulates during t
next cycle. This is a form of a <u>central</u> <u>clock</u> model in which mitosi
is part of the clock, by causing mitogen to fall, but other events o
the cycle are driven by mitosis, by one another, or by the clock. Z
then (3,9) and his colleagues (12) have performed extensive heat sho
experiments on tetrahymena which led to such a model.

The experiments we have performed on the syncytial plasmodium o
the myxomycete Physarum polycephalum during the past two years, have
convinced us that these two theories, which have dominated accounts
the cell cycle, are fundamentally inadequate, and lead to experiment
which are strategically incapable of converging on a more accurate th

believe we have begun to formulate a more adequate theory, based on
properties of sustained biochemical oscillations, to account for
llular mitotic periodicity and have already subjected our model to
vere tests.

Causal Loop Model

The view of the cell cycle as a sequence of distinct events standing
causal relations to one another, and eventually forming a loop leads
turally to the following questions: what events occur at each phase
the cycle, and what are the causal relations among these events.
assical work based on this view established the length and phase of
e DNA synthesis, S, period in a variety of organisms, and allowed the
finition of the G_1 period preceeding S, and G_2 following S and term-
ating in mitosis. We do not discuss this classical work further expt to note that in Physarum, as in some other lower eukaryotes, no
period exists. The S, G_2, M, G_1 sequence separates the cell cycle
to three phase zones. Far finer discrimination of phase markers has
come available by analysing the specific time during the cycle when
rtain enzyme activities rapidly increase in a step which is there-
ter maintained, or briefly peak, then return to a basal level. In
ysarum, about nine phase specific step and peak enzymes are known
3,14,15,47), in Saccharomyces cerevissae at least 22 are known (16,
,18,19,20,21,22) in mouse L cells seven step and peak enzymes are
own (23,24,25,26,27) and so on. The existence of such phase markers
plies some kind of cellular clock, but leaves unclear whether these
zyme markers are merely "hands" of the clock, driven by some more
ntral mechanism, or are themselves parts of the mechanism underlying
e periodicity.

One popular theory to account for the maintenance of phase rela-
ons among enzyme activities postulates a linear reading (transcribing)
genes along the chromosomes, such that spatial sequence dictates
mporal sequence. Tauro et al. (18) have compared the step timings
nine enzymes with the positions of their genes in S. cerevissae and
und them consistent with an end-to-end reading along the chromosomes.
these nine, at least four are located on the same chromosome. This
teresting possibility would account for maintenance of phase relations
ong markers on any chromosome, but not for coordination between enzymes
different chromosomes, nor for periodicity in enzyme synthesis.

III. Transition Points in the Cycle

The closed causal loop model leads naturally to search for tran
tion points between various events, or stages of the mitotic cycle.
earlier experiments in this area were accomplished with diverse meta
inhibitors or temperature shocks. Thus, a temperature shock applied
Physarum plasmodia 60 minutes before mitosis delays the subsequent m
tosis several hours (28), but the same shock applied within 7 minute
of metaphase does not delay metaphase or the subsequent completion o
the mitotic sequence and entry into S. Using actinomycin D, puromyc
and X-ray, Doida and Okada (29) have been able to establish a number
of transition points in mouse L5178Y cells: switching off of DNA sy
thesis, switching off of nuclear RNA synthesis switching on of nucle
RNA synthesis. Muldoon et al (30) showed in Physarum, using cyclohe
imide, that DNA replication itself is divisible into ten distinct ro
in which de novo protein synthesis at ten distinct times in S is req
to replicate another quantitized fraction of the nuclear DNA.

The most elegant anlysis based on the closed causal loop model
been carried out by L. Hartwell and his colleagues (11). They reaso
that the mitotic cycle can best be disected into its distinct events
collecting conditional mutants, each blocking the cell at a specific
phase of the cycle. The enormous virtues of this approach are that
allows a far finer discrimination of events in the cycle than can be
revealed by defining transition points with respect to broad classes
biochemicals such as DNA, RNA or protein synthesis; and that there i
a significant hope of establishing dependent pathways in which early
events in the cycle can be shown to be necessary (but not sufficient
conditions for the occurrence of certain subsequent events of the cy

Hartwell's analysis (11) has led him to define such a model for
the yeast division cycle. His temperature sensitive mutants define
apparent causal sequences in which a block at an earlier step blocks
subsequent events of the sequence: 1. a "start" event, initiation
DNA synthesis, continuation of DNA synthesis, medial nuclear divisio
late nuclear division, and the "start" event. 2. start, bud emerge
bud elongation, nuclear migration. Hartwell proposed that the two s
quences must join to cause cytokinesis and cell division, but that t
first sequence alone suffices to close the causal loop to initiate t
start event.

Hartwell's analysis is currently the most thorough mutant analy
of a division cycle. He has strongly favored a closed causal loop m
of the cell cycle. Nevertheless, he has two mutants which cast doub

n the simple view presented. One mutant blocks the initiation of DNA
ynthesis, an event taken to lie on the closed causal loop. Despite
his, the organism goes through repeated rhythmic rounds of bud emer-
ence. Thus, closed sequence 1 is not a necessary condition for rhy-
hmic bud emergence. Similarly, he has a mutant blocking bud emergence
hich goes through at least two rounds of S, hence the second sequence
s not necessary for the first. This data can still be accomodated
ithin a closed causal loop model; one merely postulates two separate
losed causal loops, one involving DNA synthesis, the second involving
ud emergence, and postulates that these are normally held in approp-
iate phase relations to one another. For example, budding rhythmically
ight be clocked by bud size, increasing to a threshold before initia-
ion of a new bud event. An alternate interpretation of this data is
hat there is some separate central cellular clock which ticks off
start" events and thereby drives the two separable causal sequences
hich are "downstream" of the central clock. Hartwell himself appears
o favor a hybrid model, in which the DNA-start sequence is a closed
ausal loop, while bud emergence may be timed by some separate, perhaps
entral, clock.

Despite its enormous power, the use of conditional mutants blocking
he cell at specific phases of the cycle suffers serious limitations
n any attempt to account fully for the cell cycle. The issue can be
rawn by an analogy with a grandfather clock. The weights provide the
nergy to drive the mechanism. The gears and escapements marshall the
nergy into quantitized changes occurring in the proper repeating phase
equence, the hands show the hour, and the pendulum guarantees the time
eriodicity. Without the pendulum, there may be a closed sequence of
vents which can cycle, but intervals of time are not counted, and
here is no clock. We will propose below that the cells contain an
ntity analogous to a pendulum, a sustained biochemical oscillation,
hich provides the timing of events. It is critically important that
pendulum, and a biochemical oscillation, manage to behave periodically
ue to their continuous dynamical laws of motion, not by virtue of a
losed sequence of discrete states. In terms of this analogy, mutants
f a grandfather clock stopping it at specific phases will reveal the
achinery of the hands, and of the gears and escapements, but not of the
entral pendulum. Thus, a critical limitation of conditional mutants
locking the cell at specific phases is that, by picking out such phase
specific events, it inherently overlooks continuous dynamical processes
which may be involved in timing, such as possible sustained biochemical
oscillations. We next consider the substantial evidence for such continuous timing processes.

IV. Mitogen Models

The second broad class of models to account for the periodicity
of mitosis, supposes the accumulation of a mitogen during the cycle w}
triggers mitosis at a critical concentration, is used up in the proce;
and must reaccumulate on the next cycle. Good evidence supports the
hypothesis of a cytoplasmic "mitogenic" stimulus. Most binucleate ce]
achieved by somatic cell fusion show mitotic synchrony (31,32), and s;
cytial organisms such as Physarum show startling synchrony of nuclear
mitosis. In Physarum, up to 10^{10} nuclei in a common cytoplasm divide
within two minutes of one another in a ten hour cycle. Strong evidenc
for the accumulation of a mitogen during the cycle was provided when
it was shown that fusion of a Physarum plasmodium due to undergo mito;
at 1 pm with a second plasmodium due to undergo mitosis at 3 pm led tc
their syncronous mitosis at 2 pm. If considerably more of the 1 pm
plasmodium were used than the 3 pm plasmodium, the fused pair underwer
mitosis at 1:30. Physarum strikingly demonstrates that phase, measure
as time to synchronous mitosis, behaves as though it can be averaged t
mixing a 1 pm and 3 pm cytoplasm. This strongly suggests that fusion
manages to average phases by averaging the concentrations of one or mc
continuously changing biochemical variables. On the simplest account,
the 1 pm plasmodium has a higher concentration of mitogen than the 3 p
plasmodium, and on fusion, the concentration averages out to the 2 pm
level. This data supplies the best evidence that phase behaves as tho
it were specified by the levels of one or more continuously graded var
iables like biochemical concentrations.

Weaker evidence that phase is specified by one or more substances
whose concentrations vary continuously during the cell cycle is provid
by experiments of Zeuthen and coworkers (3,4,12) using heat shocks on
tetrahymena. They found that a heat shock early in the cycle does not
delay the subsequent mitosis. As the cycle progresses, the same shock
produces a successively longer mitotic delay, until late G_2 when the
delay declines rapidly to nothing. The same form of delay curves induced by heat shocks has been found in a variety of organisms includin
Physarum (28) and higher cells (9). Such delay curves have uniformly
been interpreted to indicate the absence of mitogen synthesis early in
the cycle, the gradual accumulation of mitogen during the cycle, such
that heat shocks later produce longer delays, then the conversion of
the heat labile mitogen to a heat stable "division structure" shortly
before mitosis, to account for the drop in delay for shocks applied
shortly before mitosis. We will provide an alternate interpretation
of these results below. However, we would agree that the gradual in-

rease in mitotic delay for the same heat shock applied successively
ater in the cycle suggests the continual increase in concentration of
ome phase specifying substance, or substances, subjected to first order
estruction in the heat shock environment.

Thus cells give clear evidence both of discrete events occuring at
pecific phases of the cycle, which stand in casual relations to one
nother and also give evidence that phase can be averaged as though it
ere specified by a continuously graded "concentration" of one or more
ubstances. The simple closed loop causal model has a difficult time
ccounting for the phase averaging phenomena. The simplest mitogen
odel, with accumulation of a single variable to a threshold, is both
nlikely biochemically, and unable to account for the discrete events
appily. A more adequate general conceptual framework is required.

. Biochemical Oscillations

A start towards a more adequate framework is supplied by the large
ody of work done on circadian rhythms. A compendious review is given
y Pavlidis (33). The conceptual apparatus which has gradually come to
e employed in this field utilizes the concept of sustained oscillations
n complex, nonlinear dynamic systems. The cellular basis for such cir-
adian rhythms are unestablished, and in cases such as the eclosion
hythm in Drosophila, hard to imagine. But the similarity in phase
esetting phenomena after light shocks in many circadian rhythms, and
n mitotic delay after heat shocks in many cell systems, leads nat-
rally to the hypothesis that they are underlain by similar general
echanisms.

The possibility that biochemical oscillations might play a role
n governing the periodicity of mitosis is strongly enhanced by the
ecent discovery that such sustained biochemical oscillations do occur
n cells. The clearest example is the glycolytic oscillation in yeast,
n which that pathway can exhibit sustained spontaneous oscillations of
athway constituents, due directly to the kinetic equations linking
he synthesis and degradations of the biochemicals (34,35). Spontaneous
iochemical oscillations seem to afford one natural kind of cellular
lock, and several authors have suggested such oscillations might con-
rol mitosis (36,37). Our own simple model proposes an oscillation of
wo biochemicals, in which a threshold level in one triggers mitosis.

VI. Is Mitosis Part of the Mitotic Clock?

A natural question which arises, given any of the views of the
mitotic cycle we have described, is whether specific cell events such
as mitosis, DNA synthesis, or the occurrence of particular step or pea
enzymes which serve as phase markers are part of the clock, or are
causally downstream of the clock. In the simple causal loop model, an
event A is part of the clock if it is a member of the closed causal lo
generating periodicity, such that blocking A stops the clock, while co
accumulate at the blocked phase. If other phase markers continue to
cycle when event A is blocked, then A is not considered part of the
clock. Bud emergence is an example of the latter, as described by Har
well (11). A mutant blocking bud emergence does not block repeated
rounds of DNA synthesis.

Hartwell has shown in yeast a number of temperature sensitive mutants which do block division at specific phases, and thus cumulate ce
at that phase. This has been taken as evidence that such events are
parts of the clock in the sense that the mutants block the casual loop
and stop the clock. However, the experiments performed do not prove
that the cells' division clock has stopped at all, they may merely pro
that expression of any underlying persisting rhythmicity is blocked.
This is a critical point. Hartwell's example of a mutant blocking initiation of the DNA replication which nevertheless undergoes repeated
rhythmic budding events strongly suggests an underlying rhythmic proce
which continues despite the block in DNA synthesis, and triggers his
"start" events. Even if, in the non-permissive temperature conditions
all cells collect at some morphological stage of the cycle, an underlying clock may still be ticking independently. To our knowledge, the
is at present no convincing demonstration that any mutant blocking cel
at a specific phase actually stops the cell clock, rather than its han

In a series of interesting experiments, Mano (38,39) has obtained
striking evidence that the mitotic clock may not only be independent o
mitosis, but in some cases independent of DNA synthesis, and RNA synthesis as well. He analysed a regular cyclic variation in protein syn
thesis in a cell free system derived from the supernate of a homogenat
of clevage stage sea urchin embryos. The rhythm persisted in the pre
sence of actinomycin. The rhythm is not present in unfertilized eggs.
ATP, GTP and a maternal mRNA derived from fertilized eggs acted as
developers of the intrinsic protein synthesis rhythm in supernates of
unfertilized eggs, but did not determine the phase of that rhythm. Ma
found that the SH content of the KC1 soluble protein fraction also var
cyclically in parallel with cyclic incorporation of amino acids, and m

regulate factors determining a cyclic variation in the binding of am-
inoacyl-tRNA to ribosomes, and therefore induce a cyclic variation in
amino acid incorporation. Mano found that histone synthesis continued
cyclically in the presence of sufficient cytosine arabinoside to inhibit
DNA synthesis almost completely. In addition, tubulin was found to be
synthesized rhythmically, without DNA synthesis, but at a different phase
than histones.

Mano's experiments not only suggest that mitosis is not a necessary
part of its own clock, but fit well with the general hypothesis that the
clock consists in a sustained biochemical oscillation of many variables,
perhaps acting largely at the level of post transcriptional controls.

VII.   Clock Variables

Efforts to find the actual clock variables have been less than successful. In Physarum, Oppenheim and Katzir (40) have reported that
treatment of early G_2 plasmodia with extracts of late G_2 plasmodia phase
advanced the mitosis of the treated material. Efforts by Sachsenmaier
(41), Mohberg (42), and ourselves to repeat these experiments have not
confirmed Oppenheim and Katzir. Attempts to inject late G_2 material
into the veins of Physarum plasmodia have succeeded, but without effect
on subsequent mitosis. Technically, this is understandable, since small
volumes of material can be injected into plasmodia which must be 3 to
5 centimeters in diameter to have adequate veins, and thus represent
considerable tissue mass. Recently, Bradbury et al (43,44) have reported that phosphorylation of the F_1 histone fraction precedes chromosomal condensation, that the activity of the corresponding phosphorylating enzyme peaks in mid G_2, and that addition of exogenous calf F_1
histone phosphorylating enzyme to the surface of plasmodia in mid G_2
phase advances the subsequent mitosis.

VIII.   The Organism:

Physarum polycephalum - We have chosen to work with the lower
eukaryote myxomcete, Physarum polycephalum, having the following life
stages: A haploid spore stage which germinates to a haploid amoeba
capable of mitotic division. The amoeba can encyst as a resting haploid
state. In a liquid medium, the amoeba develops a flagellum, which it
loses on return to solid medium. Many mating types exist. Two amoeba
of appropriate mating types can fuse in syngamy to form a diploid zygote.
Homothallic mutants exist allowing an amoeba to mate with a sister. The
zygote grows by repeated nuclear division, without intervening cell

division, to form a macroplasmodium containing up to 10^{10} nuclei in a common cytoplasm. In adverse circumstances, the plasmodium follows tw pathways of differntiation; to a diploid resting sclerotial stage, or to the haploid spore stage.

The plasmodium of Physarum is among the best possible preparation for study of the mitotic cycle. Well plated plasmodia exhibit virtual complete spontaneous mitotic synchrony, in which up to 10^{10} nuclei go through metaphase within two minutes of one another during a 10 hour cell cycle. The surface plasmodium grows up to 10 to 15 centimeters in diameter. Massive nuclear cytoplasmic transplant experiments can be performed by slicing a crescent off of two plasmodia, and abutting the cut crescents. Fusion of the plasmodial membranes occurs within about 40 minutes, then exchange of cytoplasm between the two pieces is brough about by extensive cytoplasmic streaming generated in an anastomotic net of "veins" which form across the join. By labeling nuclei of one fusion partner, we have shown that on the order of 30% of the nuclei cross from one plasmodia piece to the other fusion partner in about 2 hours. Such plasmodial fusion allows massive mixing of nuclei and cyto plasm from any two well defined stages of the mitotic cycle. The mito sequence lasts about 30 minutes, and can be conveniently followed in squash preparations viewed under oil in the phase microscope.

IX.   The Oscillation of the Mitotic Clock: Limit Cycle Model

To explain the predictions we have made and tested (47,48) require describing several properties of continuous biochemical oscillations. For simplicity, we consider only our own hypothetical biochemical os- cillation. We suppose that during the Physarum mitotic cycle, a divisi protein, X, is synthesized at a constant rate A. X converts to an act- ivated form, Y, without further protein synthesis, at a rate proportior to X, BX. Y catalyzes its own formation from X at a rate proportional to X and Y^2. For example, catalytic conversion of X to Y might require activation of an enzyme by binding 2 Y molecules. Y decays at a rate proportional to its concentration. Finally, we assume a threshold leve of Y, Yc triggers mitosis.

There are no biochemical grounds to believe this particular kineti scheme is correct, and we wish to stress that at this stage of our know ledge, the importance of our model is that it is a generic example of a kinetic system capable of biochemical oscillations, which has the same critical properties as more realistic examples. We pick this particula kinetics scheme first because the feedback activation of Y on its own production parallels the feedback activation by the product of phospho-

ructokinase on the enzyme, generating the glycolytic osciallation (34).
econd, it is the simplest kinetic scheme giving rise to sustained oscil-
ations. Furthermore, it generates a wave form which characterizes the
ontrol system underlying Physarum mitotic control. In formulation,
t is rather similar to the simple division protein model which has dom-
nated the field for so long. This model was formulated prior to Brad-
ury's et al (43,44) report on the F_1 histone phosphorylation role in
ontrolling mitosis, and their contention that this biochemical is a
ood candidate for a variable of the mitotic clock. With slight alter-
tions, we could interpret F_1 histone as the X of our model, and Y,
he activated product of X, as the phosphorylated state of F_1 histone.
s we are about to show, our kinetic scheme gives rise to a sustained
scillation of X and Y concentration. A similar sustained oscillation
f the concentration of F_1 histone, and phosphorylated F_1 histone would
onstitute one possible more realistic kinetic scheme having the same
ritical general properties.

Our kinetic scheme defines a pair of differential equations linking
he synthesis and degradation of X and Y. (See Appendix)

1. $dX/dt = A - BX - XY^2$

2. $dY/dt = BX + XY^2 - Y$

The wave form of X as a function of time, and Y as a function of
ime are shown in Figure 1. The gradual increase of X, and its rather
apid drop near the time of mitosis are similar to the sawtooth wave
orm commonly assumed for the mitogen model. In Figure 2, we plot the
oncentration of Y at each moment of time, against the concentration of
at that moment. This plot gives the state of the biochemical system,
hat is, the concentrations of X and Y, at each instant of time. As
he concentrations change, the point representing the simultaneous con-
entrations in the XY "state space" changes continuously. If the system
s undergoing a sustained oscillation, it must come back to the same
tate again after one period, which will yield a closed path such as
he closed, roughly triangular path in Figure 2. As time goes forward,
he point representing the system travels counterclockwise around that
ath. A horizontal line, Yc, shows the level of Y needed to trigger
itosis. A central feature of our model is that mitosis is not part
f the mitotic clock. Were mitosis suppressed, the XY oscillation would
ontinue. Along the cyclic path, we have marked off points at successive
welfths of the total period, representing hours before mitosis.

Figure 1

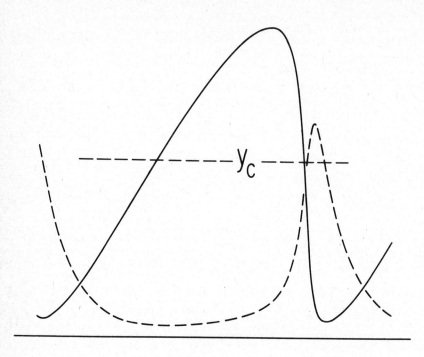

Time wave form of X and Y concentrations generated by our model mitotic oscillator for A = .5, B = .05.

If the model oscillation system is released from any specific con centrations of X and Y, or state in the XY state space, the system fol lows a unique trajectory, or path, through its state space. Since the system is deterministic, no trajectories can cross. However, some sta might not change at all with time, but might instead be a steady state A fundamental property of all real, continuous biochemical oscillation (45) is that they occur as rotations around at least 1 steady state, S. Setting equations 1 and 2 to 0, we find the steady state of our model to be Y = A, X = A/(B + A^2). If the system is released at that combina tion of concentrations, it will remain forever unaltered. This feature of all continuous oscillations is a critical distinction between our model and either the discrete loop of states model, or the familiar division protein model. Neither of these occur as rotation around a steady state. Many predictive consequences follow from this difference As noted above, we divided the cyclic path into 12 hours before mitosis

Figure 2

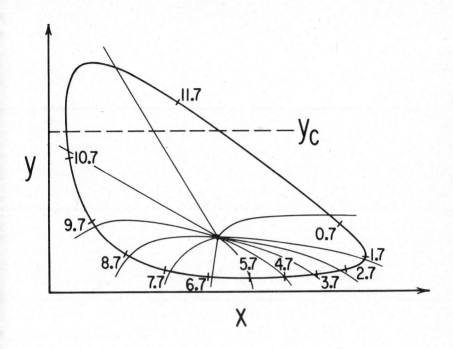

The limit cycle in the concentrations of X and Y. A critical level of Y, Yc, triggers mitosis. The nearly radial lines emanating from a point inside the cycle are "isochrons" separating equal intervals of time along trajectories, and give hours before mitosis along the limit cycle. All isochrons meet at the steady state singularity, S, inside the limit cycle.

Hence phase of the cycle is well defined on that cyclic path. But at the steady state, S, no oscillation is occurring. Therefore, S has no phase at all. Unlike the closed loop sequence of states model, or the division protein model, the biochemical oscillation model has a set of states with well defined phases, surrounding another state with no phase. In our model, if a system is released from a state near S, it follows a trajectory which spirals out to the closed cyclic path; if released from an initial state outside the closed path the system spirals onto the closed cyclic path, reaching the closed path as time goes to infinity. Such a closed path is called a limit cycle. Its critical feature is that from almost any initial condition, the system ends up on the same cyclic path, and therefore exhibits the same wave form and period biochemical oscillation. This type of stability to perturbations

is a natural prerequisite for any model of a stable periodic phenomeno
like mitosis in Physarum.

Figure 2 also shows 12 nearly radial lines emanating outward from
S, and crossing the limit cycle. These lines, or isochrones (46) separate equal 1 hour intervals of time along trajectories. Where isochrones are closely spaced along the limit cycle, angular velocity is
slow, where isochrones are far apart, angular velocity is high. An im
portant property of isochrones is that if a number of identical copies
of the oscillating system were simultaneously released from many point
on one isochrone, they would all eventually synchronize into phase
with one another as they spiral onto the limit cycle, while they continue to rotate. Thus each isochrone gives loci of equal ultimate pha
All isochrones terminate on the phaseless steady state S. Thus two
states near one another but on _opposite_ sides of S, spiral out to op-
posite sides of the limit cycle, far out of phase with one another despite being near one another and S initially. Therefore, we can see
that a small volume of states surrounding the steady state S, will spi
out to all phases on the limit cycle and therefore the small volume ha
representatives of all ultimate phases. An equal small volume of stat
centered on a point on the limit cycle, represents only a few ultimate
phases. Thus, if identical copies of our biochemical system were released from the small volume of states surrounding S, they would all
wind out to the limit cycle completely out of synchrony with one anoth
But if copies of the system were released from a small set of states
near the limit cycle, they would wind onto the cycle in near synchrony
This is a critical distinction between our, or any, oscillation model
which rotates around a set of states near S having no, or equivalently
all, phases and the classical closed loop sequence of states model, or
the division protein model, both of which are topologically equivalent
to a one dimensional closed ring of states. In the ring case, phase i
identical to state on the ring, and the ring sequence does not surroun
a phaseless state. For the biochemical oscillation phase is not identical to state on the limit cycle. All states on the same isochrone ha
the same phase. Figure 3 shows the same model, but with the numbers o
hours from each state until the first crossing of Yc, and hence the fi
triggering of mitosis. A state near the steady state S, follows a trajectory which spirals out to the limit cycle, taking several rotations
to gain sufficient amplitude to cross Yc and trigger mitosis. Therefore the state space is divisible into concentric rings of states centered on the steady state, successively 0 to 1 cycle times from crossi
Yc, 1 to 2 cycles from crossing Yc, 2 to 3 cycles from crossing Yc.
Very near S, the system is indefinitely many cycles from crossing Yc.

Figure 3

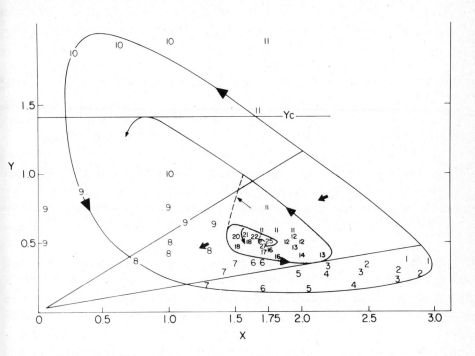

Number of hours various states of the model oscillator are from
the next mitosis. Those states which are greater than one cycle from
mitosis lie in a concentric zone surrounding the steady state (shaded
zone). The set of possible states inside the limit cycle which can be
reached with increasing durations of heat shock applied from a locus
on the limit cycle about 40 minutes prior to mitosis are shown by the
path of the solid arrows. Yc = critical threshold level of Y needed to
trigger mitosis.

. Predictions Already Tested and Confirmed

    A large set of experiments to test this model were performed using
heat shocks at 38°C. The normal growth condition is 23°C. We assume
that the dominant effect of a 38°C heat shock is to destroy X, and Y,
at a moderate rate, proportional to their concentrations. Then a heat
shock drives the system off the limit cycle, roughly toward the origin,
X = Y = 0. The effect of destroying 20% of X and Y at each phase of
the limit cycle oscillation is shown in Figure 4. The model makes a
number of striking predictions.
    1. Soon after mitosis, a heat shock drives the oscillation more

Figure 4

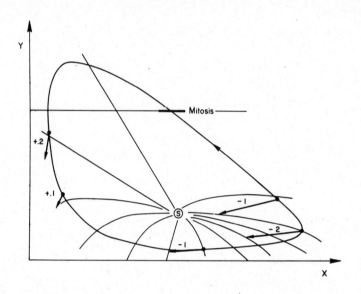

A short heat shock at any stage of the cycle, drives the system toward X = Y = 0.

rapidly in the direction it was heading, which should result in a pha advance of the subsequent mitosis.

2. After about 3 to 4 hours post mitosis, a heat shock drives system backward toward the origin, and should result in a delay. As G_2 advances the delay induced by the same heat shock should increase since the destruction "vector" is longer, representing a large retro grade movement, and because isochrones are packed closer together in the last few hours before mitosis.

3. In the last hour before mitosis, the system has "rounded th corner" and is moving up the hypotenuse of the triangular limit cycl path. A short heat shock should displace the system inside the limit cycle a short distance, but leave it on the same isochrones, from wh it should follow a trajectory to cross Yc on schedule. Hence for sh heat shocks, the delay in mitosis should increase from about 4 hours after mitosis, reach a peak about 1 hour before mitosis, then should decline to almost nothing.

4. A very striking prediction is that if a longer heat shock drives
the system further toward the origin, X = Y = 0 than does a short one,
then a longer shock just before mitosis will drive the system inside
the limit cycle, into a ring of states 1 to 2 cycles from mitosis.
Hence, as heat shock duration increases in very late G_2, mitotic delay
would suddenly jump from nearly no delay, to a bit more than a full
cycle delay. In contrast, as heat shock duration increases for shocks
applied at another phase, about 4 hours after mitosis, the system is
driven outside the limit cycle and toward the origin, rather than inside
the limit cycle. Thus, no such discontinuous jump in delays should
occur.

Figure 5A shows these predictions for 10%, 20% and 33% destruction

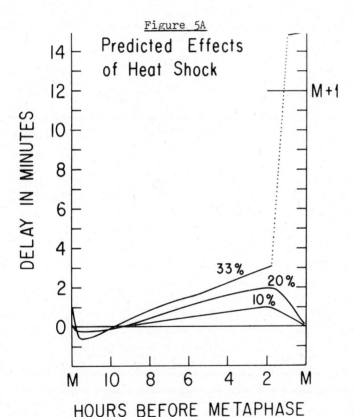

Figure 5A

Predicted Effects
of Heat Shock

Predicted effects of destroying 10%, 20%, and 33% of X and Y as a
function of phase in the cycle. Shortly before mitosis a 10% destruction
yields a slight delay, a 33% destruction yields a full cycle delay.
Shortly after mitosis destruction of 33% of X and Y causes a phase advance. Ordinate: delay in hours.

of X and Y for all phases of the oscillation. In Figure 5B, we show

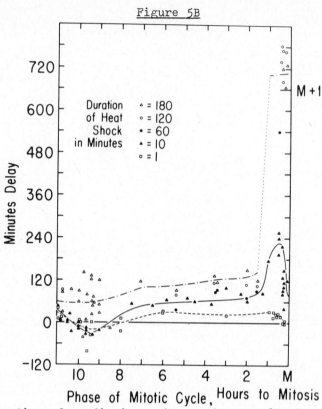

Figure 5B

Confirmation of predictions with 1 min. to 180 min. heat shocks at 38°C to Physarum. A discontinuous jump to a full cycle delay occu in late G₂ as heat shock duration increases; phase advances occur sho after mitosis.

result of a large number of heat shock experiments at all phases, in which the heat shock duration ranged from 1 to 180 minutes. We show the delay from the <u>end</u> of the heat shock, to mitosis, compared to an unshocked piece of the same plasmodium. Each point is a distinct exp iment. All predictions are confirmed. 1 and 10 minute shocks shortl after mitosis give the predicted phase advance in the next mitosis. Mitotic delay reaching a peak in late G_2, and declining to no delay f 1 and 10 minute shocks was observed. For 1, 2, and 3 hour shocks, th delay gradually increases as G_2 advances, but discontinuously jumps t a bit more than a full cycle delay as predicted in very late G_2. One of the few failures of our model is the delay induced by long heat sh shortly after mitosis, where we predict an advance. We believe that

ong shocks at this phase disrupt S, thereby inducing the unexpected
elay.

Several points are worth stressing. We have confirmed earlier
esults by Brewer and Rusch (28) showing that in Physarum, heat shocks
ive no delay early in the cycle, and give a delay increasing to a peak
n late G_2, then declining as mitosis approaches. This form of delay
urve is very widespread, having been found in Tetrahymena (3,4) Phy-
arum, and even mammalian cells (9). It has been a cornerstone for the
ivision protein model of Zeuthen, in which the gradual accumulation of
 heat labile division protein, and the late G_2 decline in delay is
aken to reflect the rapid conversion to a heat stable structure. The
bsence of delay in the early part of the cycle is uniformly interpreted
o mean no "mitogen" is synthesized at that phase. We have predicted
xactly the same form of "variable excess delay curve" directly from
he topological properties of a simple biochemical oscillation. This,
t least, casts serious doubt on the familiar interpretation. Further,
he confirmed phase advance induced by early shocks are difficult to
nterpret in the familiar division protein model. Destruction of mi-
ogen should delay mitosis. Finally, the division protein model is
ard pressed to explain why, for short shocks, as G_2 advances, delays
eaks then decline, while for long shocks delay suddenly jumps to a
ull cycle delay. The division protein model would have to assume that
he heat stable division structure is stable to short shocks, but de-
troyed all or none by long enough shocks. While possible, this is an
d hoc addition to that theory, but a predictive consequence of our
scillation model.

Our model has made a number of additional predictions which have
een confirmed.

5. After a late G_2 shock, the system is driven inside the limit
ycle, and its crossing of Yc is delayed, hence mitosis is delayed.
ut the system crosses Yc at a point _inside_ the limit cycle. From that
oint, the distance around to the next crossing of Yc is _less_ than the
ormal, full distance around the limit cycle, hence the system should
atch up on the second cycle post shock. We and others (28,48) have
onfirmed this prediction.

6. The most striking set of predictions can be seen from Figure
, which shows the concentric rings of states centered on S, success-
vely 0 to 1, 1 to 2, 2 to 3 etc. cycles from crossing Yc and under-
oing mitosis. If longer heat shocks drive the system closer to the
rgin, X = Y = 0, then in very late G_2 shocks should drive the system
nside the limit cycle, and as heat shock duration increases the bio-
hemical system should be driven further and further into the limit

cycle, crossing into zones part of a cycle from mitosis, one cycle fr&
mitosis, two cycles from mitosis, perhaps three cycles from mitosis.
Then for even <u>longer</u> shocks, the system should be driven <u>past</u> the ste&
state, S, toward the origin, back out through the rings of states unt:
it emerges in the outer ring nearest the limit cycle and the origin,
where the mitotic delay has been reduced from 2 or more full cycles, '
only about nine hours. As heat shock duration increases for very lat&
G_2 shocks, delay should first increase up to about 2 or 3 cycles, the&
dramatically decrease less than a cycle. By contrast, the same shock&
applied four hours after mitosis should result in a monotonically increasing delay, which should never be more than a full cycle, for the &
tem is driven <u>outside</u> the limit cycle, toward the orgin.

In figure 6A we show the results of many heat shocks applied for

Figure 6A

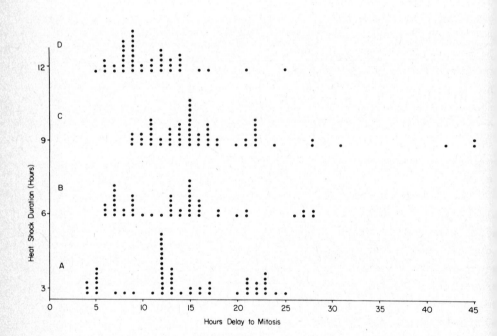

Results of 3, 6, 9 and 12 hour heat shocks applied within 40 minutes of mitosis.

3, 6, 9, or 12 hours either within 40 minutes of mitosis (late G_2), or
four hours after mitosis. All these predictions have been confirmed.

onsider first the late G_2 shocks. Each point represents only the first
ynchronous mitosis seen in one plasmidium after its heat shock. Three
our shocks resulted in delays of about 4-5 hours, 11 to 12 hours, or
2-23 hours. Thus delays of part of a cycle, one cycle, or two full
ycles are seen, for the 11 and 22 hour peaks represent successive cycle
eriods. In some of these plasmodia, no evidence of mitosis was seen
or two full cycles, then synchronous mitosis observed. Thus, it appears
hat <u>Physarum can skip mitosis for two full periods</u>, yet keep count of
ycle times up to two. This is one powerful line of evidence that mi-
osis can be suppressed in Physarum, but its clock can continue to keep
ime, and therefore that mitosis is not a necessary part of its own
lock. Further evidence supporting this will be described below.

For late G_2 shocks, as heat shock duration increases to 6 and 9
ours, delay to mitosis increases up to 2 cycles, but for 12 hours
hocks, the delay <u>decreases</u> dramatically to about 9 hours, as predicted.
hus, for late G_2 shocks, the delay has increased, then decreased as
hock duration has increased. By contrast, shocks, 4 hours after mi-
osis, show no such inversion as heat shock duration increases, Figure
B. Delays increase from 1 to 7 hours as shocks increase to 12 hours.
lasmodia do not go through mitosis in $38^{\circ}C$. We feel this is a striking
onfirmation of our general theory. It cannot be predicted at all by
he loop sequence of states model, or familiar division protein model.
ome conceivable ad hoc postulates might be appended to each however.
or example, one could say that temperature accommodation occurred after
 hours in late G_2, but never in early G_2, so in late G_2 plasmodia heated
onger than six hours, "phase" goes slowly forward, resulting in less
elay.

We tested our oscillation model in an entirely different way, by
lasmodial fusion at all possible phases and phase differences. We
odel plasmodial fusion as the fusion of two boxes, each containing a
opy of our X Y biochemical oscillation, and separated by a semipermeable
embrane which allows X and Y to diffuse between the boxes at a rate
roportional to the concentration differences. In Figure 7A and 7B,
e show the results of starting the "diffusive mixing", taken to model
umping through the anastometic net of veins across the plasmodial fus-
on join, for fusions at two different phases and phase differences.
n the first, the trajectory of the retarded plasmodium, A, crosses Yc
oughly on schedule. Its mitosis is about on time. In 7B, the initial
hase difference is larger, and A's trajectory is "tugged" inside the
imit cycle by B ahead of it, such that A curves <u>below</u> Yc, misses mi-
osis, and is subject to a long delay as it must cycle around again.

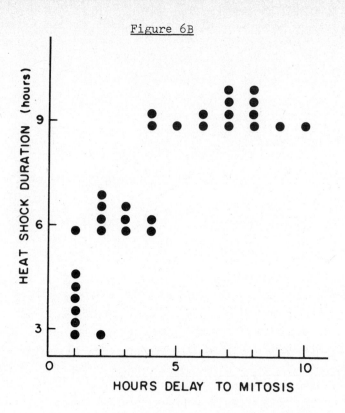

Figure 6B

Effect of 3, 6, 9 and 12 hour heat shocks applied 4 hours after mitosis.

A critical prediction, therefore, is that if the B member of the fused pair is shortly before mitosis, and A is chosen successively earlier in the cycle for fusion with such a B, then A should at first suffer little delay while its trajectory crosses Yc on schedule, then A shoul discontinuously jump to show a very long delay when it is so early wit respect to B that its trajectory curves below Yc. In Figure 8A we sho the results of many such fusions at all possible phases and phase dif- ferences. A(-) indicates no significant delay in A's mitosis. A (+) indicates a very long delay in A's mitosis, from 2 to 8 hours. For co parison, in Figure 8B, we record the predictions of our model. In bot cases, the locations of the delays, +, are identical, and both show th discontinuous jump from no delay to long delays as the phase differenc increases from less than 80 minutes to more than 100 minutes, for fusi with a B taken near its time of mitosis.

The familiar sawtooth mitogen model is incapable of fitting these results. Coupling of two such sawtooth oscillations does not even yie

Figure 7A

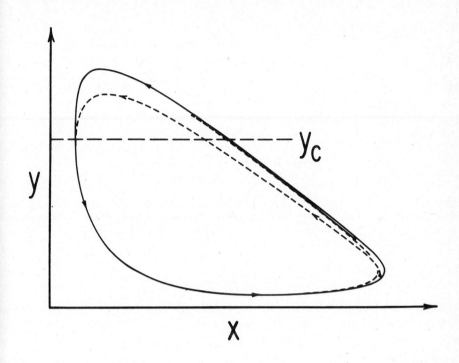

"Fusion" of two mitotic oscillators, A and B, by allowing X and Y
o diffuse between A and B, when B is 0.7 hours prior to mitosis and A
s 1.7 hours prior to mitosis. Both the trajectory from A and from B
ross Yc roughly on schedule. Both A and B go through mitosis roughly
n schedule.

ynchronization. Further, that model predicts that as phase difference
ncreases, mitotic delay increases montonically for the retarded member,
while we found no delay and a sudden jump to a long delay as the phase
difference increases. The closed loop sequence of states model has no
oherent predictions to make concerning these phenomena.

    In our fusion experiments, we discovered a new phenomenon, which we
all abortive prophase. Here nuclei enter the early stages of prophase,
but the nucleoli do not totally disperse as they normally do, rather
they break into several small clumps. No metaphase is seen, yet the
nuclei reform a larger than normal interphase nucleus and appear to
enter S, at least by the criteria of H^3 thymidimeautoradiography. We
scored these abortive prephases with triangles in Figure 8A. These
locations correspond, in Figure 8B, to fusion in which the trajectory
of A neither passed above Yc, nor clearly below it, but just grazed Yc.

<u>Figure 7B</u>

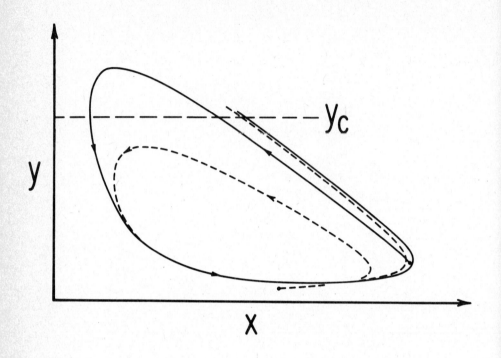

"Fusion" of B at 1.7 and A at 5.7 hours prior to mitosis. B crosses Yc but A curves below Yc and misses mitosis. Nevertheless, th mitotic oscillator continues to cycle.

We therefore suppose that abortive prephase is induced by not quite reaching threshold in Y concentration during the oscillation. In term of the interpretation of our XY oscillation as F_1 histone, and its pho phorylation, we would say that an abortive prophase is the consequence of a slightly less than adequate level of F_1 histone phosphorylation.

XI. Measuring the Waveform of the Mitotic Oscillation

We have introduced a technique to characterize the waveform of th dynamical system underlying the periodicity of mitosis in Physarum, which depends upon the fact that the synchronized phase of a fused pai of plasmodia must show certain discontinuities as the phases fused are varied systematically. Measuring a Mitotic Oscillator: The Arc Discontinuity, (47,48) describes this method. The results for Physarum strongly suggest that the dynamical system comprising the mitotic cloc

Figure 8A

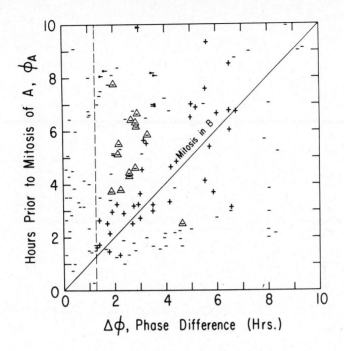

Results of plasmodial fusions on the mitosis of plasmodium A. Time before A's mitosis is plotted vertically, phase differences between A and B, horizontally. "+" means a delay in A's mitosis of two or more hours, "-" means a delay less than two hours, "△" means abortive prophase, "o-", "-o" means a dominance by one plasmodium (A and B respectively) of the phase to which the AB mix synchronizes.

...as a waveform fundamentally similar to that of our simple two variable XY oscillation model.

III. Is Mitosis Part of the Mitotic Clock?

Two lines of evidence suggest that mitosis is not itself part of the clock timing its own occurrence in Physarum. 1. Three hour heat shocks applied in late G_2, resulted in some plasmodia which skipped mitosis for two full cycles with no signs of the mitotic sequence in any nuclei. The apparent rhythmicity of the three hour late G_2 shocks results suggests the plasmodia are keeping time up to two periods with no intervening mitosis. This implies that mitosis is not a necessary part of the capacity to count two periods. It is a predicted consequence of

Figure 8B

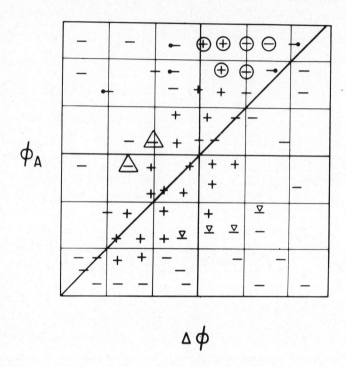

$\Delta\phi$

A phase-phase difference plot of mitotic blocks of one model mitotic oscillator when coupled by diffusion to a second model oscillato
"θ" =∅ 1 was also blocked. The other symbols, "+", "-", "Δ", "o-", an
"-o", carry the same meaning as in Figure 8A.

our model, in which mitosis is downstream of the oscillation clock.
In the same 3 hour late G_2 shocks, in some plasmodia we saw a partiall
synchronous mitosis 7 hours before the fully synchronous mitosis which
we scored in the figure. The fraction of nuclei participating in the
synchronous mitosis ranged from 5% to 90%, but had virtually no effect
on the seven hour duration until the following fully synchronous mitos.
If it were the case that passing through mitosis set the nucleus, or
cytoplasm in its vicinity to some 0 state, one would predict a longer
delay to the fully synchronous mitosis following a partial mitosis in
which 90% of the nuclei participated, than one in which 5% participated
This would be the direct prediction of the sawtooth mitogen model, but
it was not observed. This suggests that the clock is independent of
mitosis itself.

Preliminary evidence suggests that DNA synthesis is also not part

the mitotic clock. In several heat shocked plasmodia which skipped
e full cycle, we found no evidence of DNA replication by H^3 Thymidine
toradiography. We have not yet attempted to show Physarum can count
to two cycles without an intervening S period, and shall propose doing
.

II.   Evidence of Contact Inhibition of Mitosis by Plasmodial Fusion

Our detailed model, and the general concept that the mitotic clock
partially characterized as a continuous dynamical oscillation in which
threshold level of one or more variables triggers mitosis, strongly
ggests the possibility of transient or persistent subthreshold oscil-
tions induced by diffusive, or other, coupling between the oscillation
riables in adjacent cells. We model plasmodial fusions by juxtapos-
on of two identical boxes containing identical copies of the XY bio-
emical oscillation system, and allow X and Y to diffuse between copies
cording to Fick's law. We find that fusion at certain phases tends
cause both fused boxes to oscillate below the Yc threshold for two
three cycles, each box "tugging" the other inside the limit cycle,
d below Yc, Figure 9A. After a few cycles, the systems synchronize
d wind out to the limit cycle. This behavior occurs when the diffus-
n constants for X and Y are nearly equal. We found a startling be-
vior when the diffusion constants are strongly unequal, and Y is nearly
n-diffusible, while X diffused well. (See Appendix) If boxes having
milar phases were "fused", the model pair of cells synchronized rap-
ly. If two boxes far out of phase were fused, then the pair nearly
opped oscillating and the pattern of oscillation in box 1 differs from
at in box two, Figure 9B. An intuitive understanding of this pheno-
non is that when the diffusion constants of X and Y are nearly ident-
al, then at the time of fusion with box 1 in state X_1Y_1, and box 2
state X_2Y_2, the diffusive force will tend to pull each point repre-
nting one box in the XY state space, directly toward the other point
ong the line connecting them in XY state space. Hence, they tend
rongly to synchronize. But suppose only X can diffuse, not Y. Then
on fusion, the two points representing the two boxes in XY state
ace are pulled by diffusive forces parallel to one another, with no
ange in the Y dimension, as they move toward one another in the X
mension. The two points tend not to come together in XY space, and
us tend not to synchronize. The consequence of that failure will tend
be the disruption of the spatially homogeneous limit cycle oscil-
tion, and the appearance of some new pattern of stable dynamical be-
vior, such as collapse to a steady state, or a new inhomogeneous

Figure 9A

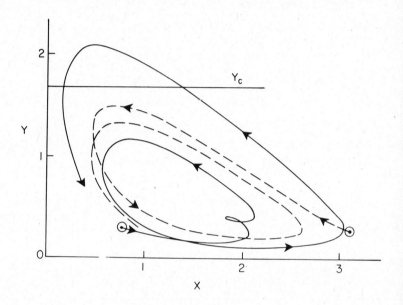

Transient subthreshold oscillations of both fusion partners ind
by fusion at the specific phases shown. This should suppress mitosi
in both fusion partners for several cycles, until the pair synchroni
and winds out to the limit cycle.

pattern of biochemical oscillation. If so, crossing of a threshold
Yc is likely to fail, and an enduring inhibition of mitosis by conta
would result.

Thus, from the simple hypothesis that the mitotic clock is part
a continuous oscillation with a threshold level of some variable tri
gering mitosis, we deduce that fusion allowing diffusive coupling (o
coupling by a variety of other "transponding" signals between non-fu
adjacent mammalian cells) can lead to transient suppression of mitos
And if some variables of the oscillation diffuse much more easily th
others, fusion is likely to cause cessation of significant amplitude
oscillations, hence induce stable contact inhibition of mitosis. A
critical consequence is that merely by effecting a gradual alteratio
in the ratio of diffusive (or other coupling) coefficients of oscill
constituents, a cell system can suddenly switch from one which tends

Figure 9B

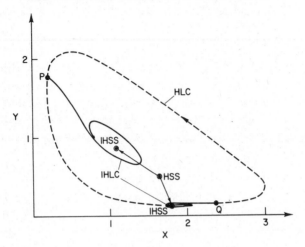

ANOTHER STABLE INHOMOGENEOUS LIMIT CYCLE.
BOXES I AND 2 ARE STARTED OUT MUCH OUT OF PHASE
(P,Q) UNDER THE SAME CONDITIONS AS FIG. 14a IN THIS
CASE THE SYSTEM EVOLVES TO A VERY CURIOUS IHLC.

Fusion induces a bifurcation of the spatially homogeneous limit
cle when the ratio of the diffusion constants of X and Y depart too
r from 1. The spatially inhomogeneous oscillation shows small ampli-
de oscillations in both fused boxes, each of inadequate amplitude to
oss Yc and induce mitosis. Hence, by tuning the ratio of diffusion
nstants, the cell can suddenly switch from one which synchronizes on
sion, to one which induces stable contact inhibition of mitosis.

nchronize and allow continued mitosis, to one which induces contact
hibition of mitosis. We think this is a very important new, simple,
d plausible general mechanism to account for contact inhibition of
owth, regardless of the particular molecules which happen to be the
nstituents of the cellular oscillation.

We already have preliminary evidence that fusions at appropriate
ases can suppress mitosis for several cycles. In some fusions in-
cing abortive prophase, normal mitosis was suppressed in both partners
r up to four full cycles.

XIV. Assessment of the Model

It seems fair to say that our model has had several predictive s
cesses. It correctly predicts that heat shocks early in the cycle pr
duce a phase advance, that for short shocks, as G_2 advances, mitotic
delay increases to a peak and declines to nothing; that long shocks ?
late G_2, but not early G_2, cause skipping of one or even two mitoses
while its clock appears to tick; that for late G_2 shocks, as the shoc
duration increases, mitotic delay first increases, then decreases dra
matically; it correctly predicts the results of very many plasmodial
fusion experiments; and preliminary evidence supports the prediction
that fusion of appropriate phases can suppress mitosis in both partne
for several cycles. In its simple form, the closed causal loop of di
crete states model of the cell cycle simply cannot make any of these
predictions. In its simplest form, the single variable mitogen model
whose threshold level of mitogen triggers mitosis and is caused by
mitosis to fall to a low level, is unable to account for these result
Nevertheless, as noted above, with sufficient ad hoc postulates it ca
fit these results, without having been able to predict them.

We would stress that we view our specific mathematical model mer
as a generic example of an oscillating biochemical system with roughl
similar properties such as wave form, period, and stability to pertur
tions. There are no compelling grounds to believe that only two var
iables like our X and Y, say F_1 histone and phosphorylated F_1 histone
are involved in timing mitosis, or that no discrete events of the mit
cycle are parts of the oscillation itself. A more adequate general v
would picture a complex web of biochemical conversions, some behaving
continuous variables, some as discrete processes which switch on and
off, all embedded in an overall oscillation which is tuned by many in
ternal parameters. This is a more creditable general view than eithe
the simple causal loop model, or the single substance, sawtooth, mito
model.

The importance of introducing the techniques of non-linear dynam
systems is that they provide a coherent conceptual framework in which
describe the _integrated_ dynamical behavior of such systems, and allow
us to devise experiments to characterize those integrated behaviors.
Non-linear dynamical systems typically partition their state spaces i
subvolumes called basins of attraction. If the system is released in
side any basin of attraction, it remains there, and flows to some asy
mptotic "attractor set"; for example, all states in the basin may flo
to a single stable steady state, or to a limit cycle. Each basin of
attraction has certain other basins as nearest neighbors in its state

ace. The importance of this is that one way to change the behavior
a dynamical system is to perturb it from one basin into a neighboring
sin, hence, it might switch from an oscillation, in the first, to a
able steady state in the second - say from the mitotic cycle to the
n-dividing G_0 state. We note that even our very simple model has two
sins of attraction: the single steady state S, is the first, and all
e rest of the state space which flows onto the limit cycle is the sec-
d. One possible location, in our oscillation theory, for the non-
viding G_0 state, would be at the non-oscillating steady state S. A
ry important feature of even our simple model is that as parameters
the XY system gradually change, first the waveform, period, and amp-
tudes of oscillations are deformed, then abrupt alterations in the
pology of the dynamical patterns of behavior of the XY system occur.
cillations appear and disappear, new steady states appear and disap-
ar. Parametric tuning is a second critical way cells might control
eir dynamical behavior. For example, in our simple model, as para-
ter A is tuned to lower values, the amplitude of the limit cycle <u>con-</u>
acts. At a critical value of A, the oscillation has contracted to a
eady state point which has now become stable. Hence, simply by tuning
parameter analogous to A, the cell could control entry into a non-
viding G_0 state. Without the concepts of nonlinear dynamical systems,
is hypothesis could hardly be stated. It is our belief, that these
tegrated dynamical patterns of behavior of the mitotic clock, and their
terations as parameters of the system are varied, constitute the ap-
ropriate <u>macroscopic biological observables</u> for the cell system, for
ese <u>are</u> the integrated behaviors of the cell.

J. Appendix

Sufficient conditions for a two-variable system of differential
quations to admit a limit cycle solution are as follows: (1) The
teady state singularity, S, must be unstable to infinitesimal perturb-
tions; vectors infinitesimally close to S must leave its vicinity.
2) From the Poincaire-Bendixon theory, the singularity must be en-
losed in a bounded domain such that on the boundary all trajectories
nter the domain.

Instability of the singularity can be shown by linearizing the
quations at the singularity, re-expressing the original equations in
erms of fluctuations away from S, and showing these perturbations
row in time.

For the model,

$$\dot{X} = A - BX - XY^2 \qquad \text{(A1)}$$

$$\dot{Y} = BX + XY^2 - Y. \qquad \text{(A2)}$$

At the steady state, $\dot{X} = \dot{Y} = 0$. Adding equations (A1) and (A2) ($\dot{X} +$ 0, yields $Y_{ss} = A$. Substituting into equation (A1)

$$\dot{X} = 0 = A - BX - XA^2$$

$$X_{ss} = \frac{A}{B + A^2}$$

The matrix of partial differentials

$$\begin{pmatrix} \dfrac{\partial \dot{x}}{\partial x} & \dfrac{\partial \dot{x}}{\partial y} \\[2mm] \dfrac{\partial \dot{y}}{\partial x} & \dfrac{\partial \dot{y}}{\partial y} \end{pmatrix} = \begin{pmatrix} -B - Y^2, & -2XY \\[2mm] B + Y^2, & 2XY - 1 \end{pmatrix}$$

may be evaluated at the steady state $Y_{ss} = A$, $X_{ss} = A/(B + A^2)$, and X and Y expressed as deviations from steady state, $X = X_{ss} + x$, $Y = Y_{ss} + y$, to yield the linearized equations in matrix form.

$$\begin{pmatrix} \dot{x} \\ \dot{y} \end{pmatrix} = \begin{bmatrix} -B - A^2, & -\dfrac{2A^2}{B + A^2} \\[4mm] B + A^2, & \dfrac{2A^2}{B + A^2} - 1 \end{bmatrix} \begin{pmatrix} x \\ y \end{pmatrix}.$$

Stability requires that the eigenvalues λ_1, λ_2 are both negative. Setting the determinant

$$\begin{vmatrix} -B - A^2 - \lambda, & -\dfrac{2A^2}{B + A^2} \\[4mm] B + A^2, & \dfrac{2A^2}{B + A^2} - 1 - \lambda \end{vmatrix} = 0$$

elds the characteristic equation:

$$\lambda^2 + \lambda \left[B + A^2 + 1 - \frac{2A^2}{B + A^2} \right] + B + A^2 = 0$$

om which

$$\lambda_1, \lambda_2 = -\frac{1}{2} \left[B + A^2 + 1 - \frac{2A^2}{B + A^2} \right] \pm$$

$$\pm \frac{1}{2} \sqrt{\left(B + A^2 + 1 - \frac{2A^2}{B + A^2} \right)^2 - 4[B + A^2]}.$$

sufficient condition for instability of S is that either λ_1 or λ_2 is
sitive. This is assured if

$$\left[B + A^2 + 1 - \frac{2A^2}{B + A^2} \right]$$

negative therefore, if $2A^2 > (B + A^2)^2 + B + A^2$. This is satisfied
r $A = \frac{1}{2}$, $B = \frac{1}{20}$. For $2A^2$ slightly greater than $(B + A^2)^2 + B + A^2$,
is oscillation is nearly linear with sinusoidal oscillations. Fixing
$= \frac{1}{2}$, and $B \rightarrow 0$, the oscillation approaches an extreme relaxation
cillator. For

$$4(B + A^2) > \left[B + A^2 + 1 - \frac{2A^2}{B + A^2} \right]^2$$

e eigenvalues are imaginary and the linearized system rotates around
e steady state. $A = \frac{1}{2}$, $B = \frac{1}{20}$ also fulfills this condition. There-
re, for these parameter values the vectors spiral out from the sing-
arity.

From the Bendixon theory to establish that this system has a limit
cle we require a bounded domain surrounding S across which all tra-
ectories flow inward toward S. Such a domain may be constructed by
tilizing the isoclines, loci of constant slope, of the vector field.
r $X = 0$, $\dot{X} > 0$, and for $Y = 0$, $\dot{Y} > 0$, hence trajctories enter across
e axes $X = 0$, $Y = 0$. The slope, S_L, of the vector field $= \dot{Y}/\dot{X}$

$$S_L = \frac{BX + XY^2 - Y}{A - BX - XY^2}$$

$$S_L = \infty \text{ if } A - BX - XY^2 = 0 \text{ or } X = \frac{A}{B + Y^2}$$

At Y = 0

$$S_L = \infty \text{ at } X = \frac{A}{B}.$$

$$S_L = -1 = \frac{BX + XY^2 - Y}{A - BX - XY^2}, \text{ or } A = Y.$$

$$S_L = 0 \text{ if } BX + XY^2 - Y = 0, \text{ or } X = \frac{Y}{B + Y^2}.$$

For $A = \frac{1}{2}$, $B = \frac{1}{20}$, the singularity is unstable to slight perturbations. For these values of A and B, the $S_L = 0$, $S_L = -1$ and $S_L = \infty$ isoclines are plotted in Figure 10. All isoclines pass through S. Since vectors enter the positive quadrant across the axes, we can def a sense to the slope -1 at X = 0, Y = A, thence to the entire vector field, which rotates around S counterclockwise. Since at Y = 0, the slope is vertical for X = 10, and negative for X > 10, one part of a Poincaire Bendixon boundary may run vertically from X = 10·2, Y = 0, to X = 10·2, Y = 0.6. For $X > A/(B + A^2) = 1·9$ and Y > 0·5, the slop of the vectors are between 0 and -1, and therefore cross into a doma whose boundary runs with slope -1 from X = 10·2, Y = 0·6 to the Y axi Near the Y axis, $Y \approx 10·5$, $X \approx 0$, vectors run into the domain with slopes between 0 and -1. This establishes the existence of a limit cycle inside the domain which rotates around S.

Figure 10

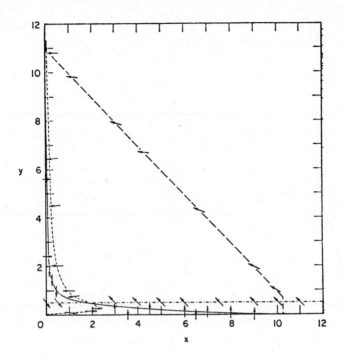

Isoclines and Poincaire-Bendixon domain establishing that equa-
ons (A1) and (A2) admit a limit cycle solution.

I. Acknowledgements

This work was a completely collaborative effort, most notably with
·s. John Wille, and John Tyson, and Mr. Carl Sheffy, at The University
· Chicago. I am deeply grateful to all three for joint experimental
.d mathematical work. The work was partially supported by NSF grant
· 36067.

II. References

. Wu, A. C., and Pardee, A. B. (1973). J. Bact. 114, 603.
. Smith, H. S., and Pardee, A. B. (1970). J. Bact. 101, 901.
. Zeuthen, E., and Rasmussen, L. (1972). Research in Protozoology
   (ed. T. T. Chen) Vol. 4, p. 9. Pergamon Press, New York.
. Zeuthen, E., Synchrony in Cell Division and Growths (ed. E. Zeu-
   then) Interscience, New York.
. Rasmussen, L. (1967). Exptl. Cell Res. 48, 132.
. Kramhoft, B., and Zeuthen, E. (1971). C. R. Lab. Carlsberg 38, 351.

7. Rusch, H. P., Sachsenmaier, W., Behrens, K., and Gruter, W. (196
   J. Cell Biol. 31, 204.
8. Sachsenmaier, W., Remy, U., and Plattner-Schobel, R. (1973) Exp
   Cell Res. 73, 41.
9. Zeuthen, E. (1973). J. Cell Sci. 13, 339.
10. Mitchison, J. M. (1971). The Biology of the Cell Cycle. Cambr
    Univ. Press, Cambridge.
11. Hartwell, L. (1974). Science 183, 46.
12. Thormar, H. (1959). C. R. Lab. Carlsberg 31, 207.
13. Sachsenmaier, W., and Ives, D. H. (1965). Biochim. Z. 343, 399-
14. Rusch, H. P. (1969). Pred. Proc. Redn. Am. Socs. Exp. Biol. 28
    1761-70.
15. Braun, R., and Behrens, K. (1969). Biochim. Biophys. Acta 195,
    87-98.
16. Sylven, B., Tobias, C. A., Malmgren, I. T., Ottoson, R., and The
    rell, B. (1959). Expt. Cell Res. 16, 75-87.
17. Tauro, P., and Halvorson, H. O. (1966). J. Bact. 92, 652-61.
18. Tauro, P., Halvorson, H. L., and Epstein, R. L. (1968). Proc.
    Acad. Sci. 59, 277-84.
19. Gorman, J., Tauro, P., La Berg, M. J., Halvorson H. (1961). Bic
    Biophys. Res. Comm. 15, 159.
20. Cottrell, S. F., and Avers, C. J. (1970). Biochem. Biophys. Res
    Comm. 38, 973-80.
21. Eckstein, H., Paduch, V., and Hilz, H. (1966). Biochem. Z. 344,
    435-45.
22. Kuenzi, M. T., and Fiechter, A. (1969). Arch. Microbiol. 64, 39
    407.
23. Littlefield, J. W. (1966). Biochim. Biophys. Acta 114, 398-403.
24. Mittermayer, C., Bosselmann, R., and Bremesskor, J. (1968). Eur
    J. Biochem. 4, 487-9.
25. Adams, R. L. P. (1969). Expt. Cell Res. 56, 55-8.
26. Gold, M., and Helleiner, C. W. (1964). Biochim. Biophs. Acta 8C
    193-203.
27. Turner, M. K., Abram, R., and Lieberman, I. (1968) J. Biol. Chem.
    243, 3725-8.
28. Brewer, E. N., and Rusch, H. P. (1965). Expt. Cell Res. 49, 79-
29. Doida, Y., and Okada, S. (1967). Nature 216, 272-3.
30. Muldoon, J., Evans, T. E., Nygaard, O., and Evans, H. (1971).
    Biochim. Biophys. Acta 247, 310-21.
31. Howard, A., and Dewey, P. L. (1960). The Cell Nucleus (ed. J. S
    Mitchell) p. 156, Betterworth, London.
32. Harris, H. (1970). Cell Fusion. Oxford Univ. Press, Oxford.
33. Pavlides, Theo. (1973). Biological Oscillators: Their Mathemat
    Analysis, Academic Press, New York.
34. Ghosh, A. K., Chance, B., Pye, E. K. (1971). Arch. Biochem. Bio
    phys. 145, 319.
35. Sel'Kov, E. E. (1968). Europ. J. Biochem. 4, 79.
36. Winfree, A. T. (1974). J. Math. Biol. 1, 73.
37. Goodwin, B. C. (1966). Nature 209, 479.
38. Mano, Y. (1968). Biochem. Biophys. Res. Comm. 33, 877.
39. Mano, Y. (1970). Develop. Biol. 22, 433.
40. Oppenheim, A., and Katzir, N. (1971). Expt. Cell Res. 68, 224-6
41. Sachaenmaier, W., Personal Communication
42. Mohberg, J., Personal Communication
43. Bradbury, E. M., Inglis, R. J., Matthews, H. R., Sarner, N. (1973
    Europ. J. Biochem. 33, 131.
44. Bradbury, E. M., Inglis, R. J., Matthews, H. R., and Langan, T.
    (1974). Molecular Basis of the Control of Mitotic Cell Division.
45. Glansdorf, P., and Prigogine, I. (1971). Structure, Stability a
    Fluctuations. Interscience Publ., Inc., New York.

6. Winfree, A. T. (1973). Biological and Biochemical Oscillators
    (ed. B. Chance, E. K. Pye, A. K. Ghosh, B. Hess). Academic Press,
    New York.
7. Kauffman, S. A. (1974). _Bull_. _Math_. _Biol_. _36_, 171.
8. Kauffman, S. A. and Wille, J. J. (1975). _J_. _Theor_. _Biol_. _55_, 47-93.

One of the most intriguing questions facing modern science is how
a proliferation of well-ordered, well-defined, but different cells can
spontaneously arise from a single cell (as, for example, how an embryo
and eventually a complete organism can arise spontaneously from a single
fertilized ovum). This question moves the mathematical models from the
familiar domain of ordinary differential equations to that of partial
differential equations; the chemistry moves from the classical domain
near-equilibrium systems to far-from-equilibrium thermodynamics. Dr.
Babloyantz presents a mathematical model based on a far-from-equilibrium
genetic system. Although her specific model is highly approximate, it
nevertheless provides a physio-chemical basis for positional information
amongst cells which could lead to pattern formation and morphogenesis.
If models like the one discussed here by Dr. Babloyantz are correct,
they would be particularly useful to biologists because they relate
intuitively obvious physical and chemical events (diffusion and biochemical reactions) to well-known biological events (cellular differentiation).

MATHEMATICAL MODELS FOR MORPHOGENESIS

A. Babloyantz,
Université Libre de Bruxelles,
Faculté des Sciences,
1050 Bruxelles, Belgium.

. Introduction.

Open chemical systems exhibiting nonlinear kinetics and at a finite distance from thermodynamic equilibrium may evolve spontaneously due to the action of a random disturbance on a time or space dependent regime Glansdorff and Prigogine, 1971). For a particular class of kinetic equations, multiple steady states may also appear (Babloyantz and Nicolis, 972). These organizational processes can only be created or maintained y dissipative entropy-producing processes inside the system. A gen-ralized thermodynamics has been developed for these "dissipative struc-ures" by Prigogine and Glansdorff (1971). Nicolis and Auchmuty (1974) ave shown their existence analytically and confirmed the wide variety f properties of these structures obtained by Herschkowitz-Kaufman and Nicolis (1972) and Nicolis (1974a) in computer simulations.

It seems that these ordering phenomena may provide a physico-chem-cal basis for the understanding of certain biological systems. Indeed, he latter are highly complex and ordered objects, operating far from chermodynamic equilibrium. They are subject to induction, repression, cross-feedback, and cooperative mechanisms which obey nonlinear kinetics. Moreover the morphological patterns of living systems are characterized by highly heterogeneous distribution of matter which results from the processes of embryological development.

Computer studies of model systems which may give rise to dissipative structures show a wide variety of spatial and temporal behaviour. We report here briefly some of the more salient properties observed.

For fixed boundary conditions it is seen that when structure for-mation is possible, the steady state corresponding to the extrapolation of the close to equilibrium behaviour becomes unstable. Then the homo-geneous system evolves to a new regime corresponding to

1) inhomogeneous distribution of chemical substances in space.
2) Under different conditions the concentration of the chemicals may show sustained oscillations (Nicolis and Portnow, 1973).

Such oscillations have been found in glycolysis; Goldbeter an
Lefever (1972) have recently developed a model for these osci
lations.

3) One can also obtain a regime depending on time and space whic
   is characterized by propagating concentration waves (Herschkowitz-Kaufman and Nicolis, 1972). This may be relevant to
   the description of information propagation as in the case of
   aggregating slime molds (Robertson and Cohen, 1972).

4) Multiple steady states and hysteresis may appear for certain
   type of kinetics. This kind of behaviour can be used for the
   description of biological phenomena exhibiting "all or none"
   type models for genetic regulation (Babloyantz and Nicolis,
   1972), β-galactosidase induction in E. Coli (Babloyantz and
   Sanglier, 1972), excitable systems (Blumenthal et al., 1972)
   and the synthesis of prebiotic polymers (Goldbeter and Nicoli
   1972; Babloyantz, 1972; Prigogine et al., 1972).

Spatial structure formation and wave propagation can only appear
if the appropriate nonlinear chemical reactions proceeding in differen
domains of space are coupled by diffusion phenomena.

Recently we (Babloyantz, unpublished) have found that other opera
tors which couple different portions of space may give rise to dissipa
tive structures. An ensemble of discrete chemical systems in contact
are a good example of such systems. This later property enables one t
apply the theory of dissipative structure to a still wider range of
biological phenomena such as neural networks.

Now if instead of fixed boundary conditions one assumes that the
flux of diffusing substances are zero across the boundaries of the sys
the new type of dissipative structures illustrated in Figure 1 may app
and give rise to gradient formation in the system.

In what follows we shall try to apply some of the properties of
dissipative structures such as gradient formation and appearance of
multiple steady states to the study of a specific biological system:
namely the problem of generation of polarity and pattern formation
in an initially homogeneous morphogenetic field.

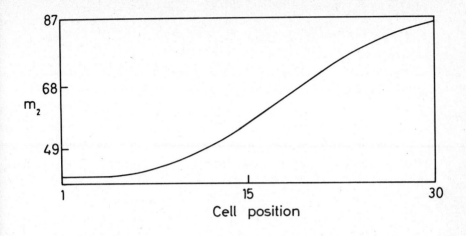

<u>Figure 1</u>: Concentration profile of m_2 for L = 0.9.

II. <u>Models for morphogenesis</u>.

A living system is formed by an ensemble of differentiated cells arranged in space according to a given and well-defined order. Thus, one of the major problems of developmental biology is to understand how a given genome, identical in all cells of a multicellular system, can be expressed to give specific and varying spatial patterns of cellular differentiation.

The existing theories of pattern formation postulate a universal two-step process with two distinct time scales (See for example, Wolpert, 1969 and Crick, 1970.). In a first stage, positional information is assigned to the cells in a coordinate system by a gradient generating mechanisms. In a second stage, the action of positional information at the genetic level gives rise to molecular differentiation. However all these models are based on the assumption of preexisting inhomogeneities in the system (Gierer and Meinhardt, 1972).

Here we shall outline an approach where starting from an ensemble of identical cells (i.e. a homogeneous morphogenetic field), a gradient

can be generated in the system spontaneously. The latter acting at th
genetic level gives rise to cellular differentiation and pattern formation (Babloyantz and Hiernaux, 1975).

Before reporting the rather involved model exhibiting the above
properties, it is useful to separate the problem into two parts. In
the first part we show how a gradient can be generated in a homogeneou
system. In the second stage we assume the existence of the gradient a
show how positional differentiation may occur. The complete model is
discussed in a later section.

III. Gradient formation-polarity generation.

Any coherent theory of pattern formation must explain how polarit
may be induced and maintained in an initially uniform tissue. The
conditions for the onset of such inhomogeneities was first investigate
by Turing (1952) who showed that a system of chemical substances react
together and diffusing through a tissue may describe the phenomena of
morphogenesis. In his original paper, Turing did not consider the sta
bility of his patterns.

However now, in the general framework of "dissipative structures"
we can show that gradient formation is the result of the first bifurcation of the steady state of a set of reaction-diffusion equations,
providing zero flux boundary conditions are imposed on the system. We
consider the following chemical reactions (see Babloyantz and Hiernaux
1975):

$$a \xrightarrow{k_1} m_1$$

$$c + 2m_1 \xrightarrow{k_2} m_2 + 2m_1$$

$$m_1 \xrightarrow{k_3} f$$

Scheme 1.

$$m_2 \xrightarrow{k_4} g$$

$$m_2 + p \underset{k_6}{\overset{k_5}{\rightleftarrows}} m_1$$

$$b + m_1 + p \xrightarrow{k_7} p + 2m_1$$

a is the precursor of morphogen m_1. The latter in turn catalyzes the formation of another morphogen m_2. This morphogen will combine with a third molecule p to give m_1. The molecule p also catalyzes the formation of m_1. Finally, m_1, and m_2 may decay. $k_1 \ldots k_7$ are the kinetic constants. b and c are respectively precursors of m_1 and m_2. a, b, c, f and g are maintained space- and time-independent throughout the system. Assuming that the time variation of p is quasi-stationary the kinetic equations describing the system are:

$$\frac{\partial m_1}{\partial t} = k_1 a + \frac{k_6 k_7}{k_5} b \frac{m_1^2}{m_2} - k_3 m_1 + D_{m_1} \frac{\partial^2 m_1}{\partial r^2}$$

$$(1)$$

$$\frac{\partial m_2}{\partial t} = k_2 c m_1^2 - k_4 m_2 + D_{m_2} \frac{\partial^2 m_2}{\partial r^2}$$

where D_{m_1} and D_{m_2} are diffusion coefficients.

It has been shown (Babloyantz and Hiernaux, 1975) that there exists a time-independent homogenous solution for this set of equations. However for a certain range of parameters and values of D_{m_1}, D_{m_2} these steady states are unstable and the system evolves to a new solution. The nature of the latter is entirely dependent on the values of parameters and also on the dimension of the morphogenetic field. One can obtain spatial structure formation, wave propagation and homogeneous oscillatory behaviour.

Figure 1 shows a structure obtained with a set of parameters for which a space-dependent steady state is formed. The length of the field is such that it corresponds to the first bifurcating solution of the homogeneous steady state; a theorem in bifurcation theory asserts the stability of this pattern (Sattinger, 1973). If the length of the system is increased the number of possible bifurcating solutions increases and more complex patterns with several high and low concentration regions appears. Thus growth engenders a whole succession of forms and patterns.

IV. Models for positional differentiation.

In this section we assume that by a mechanism such as the one described in the preceding paragraph a gradient is already established in the morphogenetic field. We investigate here the cellular differentiation and pattern formation of an ensemble of N initially homogeneous cells in this field.

The cells of a morphogenetic field are in contact, and this fact enables a morphogen to spread throughout the field. One can assume di ferent mechanisms for the passage of a substance from cell to cell:

i) Cell communication via membrane receptors (McMahon, 1975).

ii) Diffusion of the morphogen throughout the field (Babloyantz and Hiernaux, 1975).

iii) Active transport of morphogen from cell to cell (Babloyantz a Hiernaux, 1974).

We shall be considering here only the last two phenomena which can be incorporated in the following model.

In this model the diffusion of a morphogen S responsible for the establishment of positional information at the macroscopic level is coupled to an induction mechanism for enzyme synthesis at the cellular level.

The morphogenetic field is formed by a linear array of N identica interacting cells subject to a source and a sink of morphogen at the boundaries of the field. Each cell performs the following chemical reactions:

$$R' \xrightleftharpoons[h_2]{h_1} R_i$$

$$R_i + O_i^+ \xrightleftharpoons[h_4]{h_3} O_i^-$$

$$R_i + 2I_i \xrightleftharpoons[h_6]{h_5} F_1$$

$$\alpha + O_i^+ \xrightarrow{h_7} E_i + O^+$$

$$S_i + E_i \xrightleftharpoons[h_9]{h_8} I_i + E_i$$

$$E_i \xrightleftharpoons[h_{11}]{h_{10}} F$$

Scheme II

A repressor R is synthesized from its precursors R' and can repres the operator O^+ into O^-, thus blocking the synthesis of the enzyme E. The latter catalyzes a reaction whose substrate is the morphogen S and

hose product I acts as an inducer by inactivating the repressor R.
$_1 \cdots h_{11}$ are the rate constants, F is the decay product of the enzyme
, α represents the precursors of enzyme E and F_1 is the repressor-
nducer complex.

The steady states of the kinetic equations of Scheme II have the
roperty of exhibiting an "all or none" type of behaviour after a given
ell reaches a critical concentration S_c. At this value the concentra-
ion of enzyme E suddenly jumps from a negligible quantity to a much
igher value. When this happens, assuming that E is a "luxury" protein
e say that the cell is differentiated as opposed to the undifferentiated
tate before the transition occurs.

The above-cited chemical reactions are coupled throughout the entire
ield by diffusion of S from cell to cell. We assume that the later
rocess can be approximated by Fick's law and moreover, is written in a
inite difference approximation suitable for numerical calculations. The
inetic equations describing the time change of concentrations in the
ield are given by:

$$\frac{\partial E_i}{\partial t} = \frac{\alpha h_2 h_4 h_7 + \alpha h_4 h_5 h_7 (I_i)^2}{h_1 h_3 R' + h_3 h_6 F_1 + h_2 h_4 + h_4 h_5 (I_i)^2} - h_{10} E_i + h_{11} F$$

$$\frac{\partial I_i}{\partial t} = -2h_5 \frac{h_1 R' (I_i)^2 + h_6 F_1 (I_i)^2}{h_2 + h_5 (I_i)^2} + 2h_6 F_1 + h_8 E_i S_i - h_9 E_i I_i \qquad (2)$$

$$\frac{\partial S_i}{\partial t} = - h_8 E_i S_i + h_9 E_i I_i + \frac{D_s}{r^2} (S_{i+1} + S_{i-1} - 2S_i)$$

$$\text{for} \quad i = 2, \; N-1$$

$_s$ is the constant diffusion coefficient and r is the dimension of a
ell.

This set of equations have been integrated by computer simulations
o provide a steady state solution. The boundary cells 1 and N are
;iven values corresponding respectively to that of E before S_c and after
he "all or none" transition threshold of a single cell. The low value
orresponds to the sink and the high value is that of the source.

Figures 2 and 3 represents respectively the values of S and E in
he field. The position of the source determines the polarity of the

field.   We see that a smooth variation in the gradient of morphogen ma
generate a quasi-discontinuous response curve for E.   In the morphoger
tic field of N cells due to the process of diffusion, the concentrati
of S is not identical in all cells.   The ones with concertation $S > S_c$
synthesize appreciable quantities of E and are differentiated and thos
with $S < S_c$ remain undifferentiated.

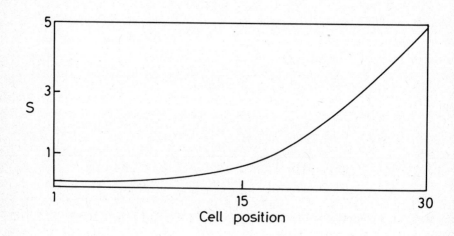

<u>Figure 2</u>:   Gradient of diffusion of morphogen S in a source-sink field
of 30 cells.

For a given dimension of the field, the proportion of induced cel
is determined by the value of the concentration of S at the source and
at the sink and the magnitude of diffusion coefficient D_s.

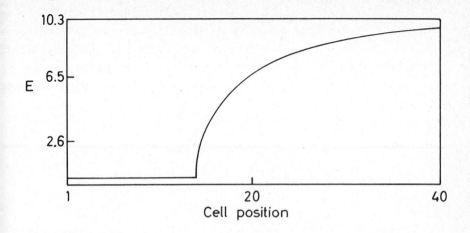

Figure 3: Pattern of cell differentiation corresponding to the gradient of Figure 2.

. Positional differentiation and active transport.

The establishment of a gradient of "positional information" may also result from an active transport of morphogen between a source and sink throughout the field of N cells. Each cell is performing the same type of protein synthesis as the one of the preceding paragraph. However, inducer is transported into the cells by the action of a permease (See Babloyantz and Hiernaux, 1974). Different chemical reactions of the field are

$$R' \underset{k_2}{\overset{k_1}{\rightleftarrows}} R_i$$

$$O_i^+ + R_i \underset{k_4}{\overset{k_3}{\rightleftarrows}} O_i^-$$

$$R_i + 2I_i \xrightleftharpoons[k_6]{k_5} F_1$$

$$\alpha + O_i^+ \xrightarrow{k_7} O_i^+ + E_i + M_i$$

$$M_i + I_{i-1} \xrightleftharpoons[k_9]{k_8} M_i + I_i$$

Scheme III.

$$M_i + I_{i+1} \xrightleftharpoons[k_9]{k_8} M_i + I_i$$

$$M_i \xrightleftharpoons[k_{11}]{k_{10}} F$$

$$E_i \xrightleftharpoons[k_{13}]{k_{12}} G$$

where M is the permease synthetized in the same time as E.

It must be noticed that the equations for each cell are symmetrical. The only dissymmetries are the source and sink values at the bou aries. It has been shown (Babloyantz and Sanglier, 1972) that if a single cell synthetizing permease is embedded in a solution containing inducer I, there is a threshold value I_c for which, as in the precedin case, the cell exhibits all or none effects with respect to protein syntheses and begins to perform active transport. As in the preceding case, the source and sink values correspond to the concentration of E before and after the transition value of I.

The kinetic equations describing this system are

$$\frac{\partial E_i}{\partial t} = \frac{\alpha k_2 k_4 k_7 + \alpha k_4 k_5 k_7 I_i^2}{k_1 k_3 R' + k_3 k_6 F_1 + k_2 k_4 + k_4 k_5 I_i^2} - k_{12} E_i + k_{13} G$$

$$(3)$$

$$\frac{\partial M_i}{\partial t} = \frac{\alpha k_2 k_4 k_7 + \alpha k_4 k_5 k_7 I_i^2}{k_1 k_3 R' + k_3 k_6 F_1 + k_2 k_4 + k_4 k_5 I_i^2} - k_{10} M_i + k_{11} F$$

$$\frac{\partial I_i}{\partial t} = -2k_5 \frac{k_1 R' I_i^2 + k_6 F_1 I_i^2}{k_2 + k_5 I_i^2} + 2k_6 F_1 + k_8 M_i (I_{i-1} + I_{i+1})$$

$$- 2k_9 M_i I_i - k_8 I_i (M_{i-1} + M_{i+1}) + k_9 M_{i-1} I_{i-1} + k_9 M_{i+1} I_{i+1}$$

$$k = 2, \ldots, N - 1$$

Figure 4 represents the steady state solution of these equations for E starting from a state where the 40 cells are undifferentiated. It is seen that the curve for synthesis of E is quasi-discontinuous and there is a net separation among the cells of the field into two categories. One part of the cells remains undifferentiated and is unable to perform active transport. The other part on the contrary is differentiated and transports I actively from regions of low I to regions of high I concentration.

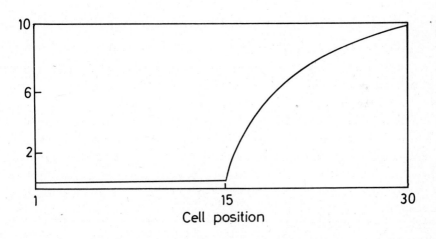

Figure 4: Pattern of cell differentiation for a source-sink active transport field of 40 cells.

An important property of these patterns is that the value of I a[t] the source organizes the pattern. Indeed, for a fixed source value th[e] number of cells that are differentiated remains constant whether one adds or substracts several cells from the morphogenetic field. Thus positional information "flows" from source to sink. This defines the polarity of the pattern.

VI. Self-consistent model for positional differentiation.

In this section we present a model that exhibits the different properties of sections IV and V, and thus enables us to have in the same model gradient formation and positional differentiation.

In order to do this let us suppose that two substances m_1 and m_2 react according to Scheme II. Then m_2 gives rise to a substance I whi[ch] in turn is transported actively from cell to cell and acts on the cell genome according to Scheme III.

Again assuming all the variables at their steady state but the c[on]centrations m_{1i}, m_{2i}, E_i, M_i and I_i, the time evolution of this system is given by:

$$\frac{\partial m_{1i}}{\partial t} = k_1 a + \frac{k_6 k_7}{k_5} b \frac{m_{1i}^2}{m_{2i}} - k_3 m_{1i} + \frac{D_{m_1}}{r^2}(m_{1i-1} + m_{1i-1} - 2m_{1i})$$

$$\frac{\partial m_{2i}}{\partial t} = k_2 c m_{1i}^2 - k_4 m_{2i} + \frac{D_{m_2}}{r^2}(m_{2i-1} + m_{2i+1} - 2m_{2i})$$

$$\frac{\partial E_i}{\partial t} = \frac{\alpha h_2 h_4 h_7 + \alpha h_4 h_5 h_7 I_i^2}{h_1 h_3 R' + h_3 h_6 F_1 + h_2 h_4 + h_4 h_5 I_i^2} - h_{10} E_i + h_{11} F$$

$$\frac{\partial M_i}{\partial t} = \frac{\alpha h_2 h_4 h_7 + \alpha h_4 h_5 h_7 I_i^2}{h_1 h_3 R' + h_3 h_6 F_1 + h_2 h_4 + h_4 h_5 I_i^2} - h_{12} M_i + h_{13} G$$

$$\frac{\partial I_i}{\partial t} = -2 \frac{h_1 h_5 R' I_i^2 + h_5 h_6 F_1 I_i^2}{h_2 + h_5 I_i^2} + 2h_6 F_1 + h_8 M_i(I_{i-1} + I_{i+1})$$

$$- 2 h_9 M_i I_i - h_8 I_i (M_{i-1} + M_{i+1})$$

$$+ h_9 M_{i-1} I_{i-1} + h_9 I_{i+1} M_{i+1} + w\, m_{2i} \qquad k = 2,\ldots,N-1$$

(4[ ])

For the chosen values of parameters of the system the time evolution
f the set of equations (4) shows first the appearance of a dissipative
tructure of the same type as the one of Figure 1, giving rise to a
ource and a sink in concentrations of m_1 and m_2. The values of the
hree other variables remain constant. Once the gradient is established
he active transport of I from cell to cell may proceed and give rise
o cellular differentiation and pattern formation of the type seen in
igure 4.

The foregoing paragraph shows that the dissipative structures in-
uced by the flow of free energy under non-equilibrium conditions and
onlinear kinetics give rise to a new physical chemistry most suitable
or the understanding of emergence of biological order.

II. <u>References</u>.

abloyantz, A. (1972). <u>Biopolymers</u> 11, 2349.
abloyantz, A. and Hiernaux, J. (1974). <u>Proc. Natl. Acad. Sci.</u> 71,
530-1533.
abloyantz, A. and Hiernaux, J. (1975). <u>Bull. Math. Biol.</u> 37, 637, 659.
abloyantz, A. and Nicolis, G. (1972). <u>J. Theor. Biology</u> 34, 185-192.
abloyantz, A. and Sanglier, M. (1972). <u>FEBS Letters</u> 23, 364.
lumenthal, R., Changeux, J. P. and Lefever, R. (1970). <u>J. Membrane</u> 2,
71.
rick, F. (1970). <u>Nature</u> 225, 420-422.
ierer, A. and Meinhardt, H. (1972). <u>Kybernetik</u> 12, 30-39.
lansdorff, P. and Prigogine I. (1971). "Thermodynamics of Structure,
tability and Fluctuations", New York, Wiley-Interscience.
oldbeter, A. and Lefever R. (1972). <u>Biophys. J.</u> 12, 1302.
oldbeter, A. and Nicolis, G. (1972). <u>Biophysik</u> 9, 212.
erschkowitz-Kaufman, M. and Nicolis, G. (1972). <u>J. Chem. Phys.</u> 56, 1890.
cMahon, D. (1975). <u>Proc. Nat. Acad. Sci. U.S.A.</u> 70, 2396-2400.
icolis, G. (1974). <u>SIAM-AMS Proceedings</u> 8, 33.
icolis, G. and Auchmuty, J.F.G. (1974). <u>Proc. Nat. Acad. Sci. U.S.A.</u> 71,
748-2751.
icolis, G. and Portnow, J. (1973). <u>J. Chem. Rev.</u> 73, 365.
rigogine, I., Nicolis, G. and Babloyantz, A. (1972). <u>Physics Today</u> 25,
° 11 and 12.
obertson, A. and Cohen, M. H. (1972). <u>Ann. Rev. Biophys. Bioengineering</u>
, 409.
attinger, D. (1973). "Lecture Notes in Math." 309 Berlin, Springer.
uring, A. M. (1952). <u>Phil. Trans. Roy. Soc.</u> B237, 37-72.
olpert, L. (1969). <u>J. Theor. Biol.</u> 25, 1.

The hierarchial step from subcellular biochemical systems to mult
cellular physiological systems is sufficiently immense to boggle the
most brilliant, even the simplest mind. There are those who would arg
that trying to describe physiological systems in terms of the constitu
biochemical variables is like trying to describe the behavior of air-
planes in terms of the dynamics of the constituent atoms and molecules
Nevertheless, there is a relationship; the question is whether or not
it will be observable in the noise of a real world. Professor Garfink
argues convincingly with concrete examples that physiological systems
can indeed be modeled in terms of a finite number of relevant biochemi
variables, and that the potential contributions of such models to discoveries about human biology and medicine justify the effort necessary
to overcome the stragetic as well as the tatical problems involved.

COMPUTER SIMULATION AS A MEANS OF PHYSIOLOGICAL INTEGRATION OF BIOCHEMICAL SYSTEMS

David Garfinkel, Lillian Garfinkel, and
William Terrell Moore
Moore School of Electrical Engineering
University of Pennsylvania
Philadelphia, PA   19174

Introduction

Traditionally biochemistry has been concerned with analyzing the components of living systems, and has paid less attention to how they interact. While the information available about individual entities such as proteins and nucleic acids is now immense and still growing rapidly, the task of determining how they interact to form a living system is now only beginning, especially with respect to quantitative behavior. In reviewing a recent book, Dr. J. S. King (1975), the editor of Clinical Chemistry has written "biochemistry was mainly a reductionist science; only lately has it become possible to fit together facts and observations into what is beginning to be a coherent -- and thus more interesting -- whole".

Here we describe the contribution that the modeling of biochemical systems can make to this process. The models described here are concentrated in those areas of biochemistry where the amount of quantitative data is the greatest: the metabolism by which metabolic fuel is converted to energy, specifically glycolysis, the Krebs cycle, and related metabolism. Examples from outside this area will mostly serve as counter-examples. The technology for building such models can be only briefly reviewed here, and is being described in more detail elsewhere (Garfinkel et al., 1976a).

As we are concerned with the combination of biochemical information to produce physiological results, some indication seems necessary of where and how ill-defined the boundary between these two disciplines is. It is clear that proteins are biochemical entities, that organs are at the physiological level, and that membranes and cellular substructures are somewhere in between and may

be studied from either point of view. Much of the work considered her
is in this in-between area; model input is primarily biochemical where
the output obtained is more physiological. To put it another way, muc
of the work here described is at or near the boundary between two hier
archical levels of biological structure, biochemistry and physiology.

Brief Review of the Technology

The very simplest biochemical models, those involving a few simpl
individual enzymes, can be manipulated with pencil and paper, or at mo
a programmable desk calculator (e.g., Rapoport et al., 1976). Compute
are needed to manipulate and analyze more complex models. Specialized
computer techniques applying numerical analysis, and other relevant di
ciplines have been developed for these cases. The complexity of most
physiological models tends to be intermediate between the two above ex
tremes, but the most complex physiological models (Guyton et al., 1972
resemble bidochemical counterparts in their technical requirements.

We may divide biochemical models according to whether or not time
is a variable; the latter are simpler. Again, single-enzyme models ar
the simplest representatives of this class, and computer manipulation
with relatively simple programs is feasible. We have recently reviewe
this subject elsewhere (Garfinkel et al., 1976b). At the more complex
end of this scale are the elaborate chemical-equilibrium calculating
programs such as those developed by the RAND Corp., (DeLand, 1967) whi
permit calculating the final equilibrium composition of a chemical mix
ture from the initial composition and the known equilibrium constants
or free energies of the components.

Commonly biochemical and physiological models of any complexity d
have time as their independent variable. Computers are required for
such models which consist of a set of simultaneous differential equati
which are normally too complex for pencil and paper solution. Analog
computers are sufficient for the simpler physiological models, while d
ital computers with their larger capacities are needed for most biochemical as well as the more complex physiological models.

It is a common finding that the differential equations which comprise these models are "stiff"; they are numerically badly behaved so
that their solution is very slow and expensive. This problem, which
has received some serious attention from the numerical computing community since 1968, has by now been considerably alleviated if not completely solved. A moderately complex glycolysis model (Garfinkel and
Hess, 1964) which cost $2,000 to compute through the first time in

962, has by now had its computation cost reduced by three orders
f magnitude and there is more improvement pending. The most im-
ortant factor here is the application of the algorithm of Gear (1971)
or solving "stiff" equations by an implicit variable-order method.
e have been able to improve this further for biochemical models so
hat it requires both less time and less core memory storage (Roman
t al., 1976).

Although it is possible to write these differential equations
n a general computer language like FORTRAN, their complexity makes
he use of some specialized simulation language desirable. We have
een working with a language that contains many useful features
dded over its effective 20-year history (Garfinkel, 1968, and
evisions in preparation.) Basically this language generates dif-
erential equations from chemical equations. Input to this program
s in the form of chemical equations, with the values of rate con-
tants and chemical concentrations being specified. When the number
f differential equations rises to the hundreds, the kind of assist-
nce which these languages provide in monitoring model behavior is
ecessary. The computer can be applied to other phases of this work
s well; we sometimes feel that we are working at the limits of the
uman memory in trying to keep track of all the biochemical infor-
ation that must go into a model. Ultimately computer data-banding
echniques should be applied here, but considerable technical develop-
ent is needed before this is possible.

When the technical problems of writing and solving differential
quations have been overcome, the modeler is concerned with deter-
ining which model (the set of differential equations and its associ-
ted numerical parameters) adequately fits the biochemical data.
his problem usually is solved by applying optimization techniques,
hich select that set of parameters for a given set of differential
quations which minimizes the sum of the squares of deviations
etween computed and observed behavior. The application of these
echniques to physiological models has recently been reviewed by
ohnson (1974). Other considerations for when a model fits expermental data properly have been developed by Reich (1970, 1974) and
eviewed by us (Garfinkel et al., 1976b). The most important limit-
tions here are that optimizing many variables at once causes dif-
iculty, as does the presence of relationships among them (constrained
ptimization). To some extent this may be overcome by building a
odel in pieces, which are optimized separately and then combined;
then one need only optimize the variables which interrelate the pieces.

To some extent the choice of what models or differential equations to try has depended on intuition, especially for the more complex situations. Formal techniques such as the SAAM program of Berman (1965) which fits data on compartmental systems to models from a library are available for these specialized situations, although they are slow in operation. While some other standard techniques of systems analysis, such as sensitivity analysis, may be helpful, it may be necessary to utilize advanced computer-based techniques, such as artificial intelligence, to speed the process of model development from its present unreasonable slow state.

Data Requirements

It is not completely clear how much input biochemical data are needed to determine a complex model, since biological systems (which often give the impression of having been designed by the late cartoonist Rube Goldberg) are very highly interconnected. Thus an error in one parameter will be propagated or shared throughout a complex system and it is likely to be caught because it is likely to contradict something else. Or inarily one wants considerably more input measurements than the number of variable model parameters so that one can choose the best model by optimization techniques. In actual fact the amount of data required to determine a complex metabolic system in this way may considerably exceed what can be obtained at one time from one preparation, so that the available data on the entire system may be of differeing kinds from different sources. It is then necessary to combine them with data on isolated enzymes and other subunits of the system being studied. A sufficiency of data may not be enough: Raugi et al. (1975) found that having enough input data was nevertheless not sufficient to uniquely define their model of metabolism in _Tetrahymena_. They determined more than enough data points, so that their model was overdetermined, but they nevertheless were not able to determine a unique model -- and could not improve the fit of their model by introducing additional parameters into it.

A number of guidelines emerge from our experience with transient changes in multi-enzyme pathways whose component enzymes are well characterized. In such cases we need to know what goes into the pathways, what comes out of it (including the rate of oxidation of organic molecules to CO_2), the time course of at least one intermediate substance in the pathway, and the initial concentractions of all the significant intermediates. One would like to be able to

eparate biological variability and analytical error as far as pos-
ible, as one animal may consistently have higher values for every-
hing than another, and this may be lost when pooling data.

The modeler generally wants access to what was actually measured
y the experimenter. The process of editing that accompanies pub-
ication is not helpful here, as it emphasizes conclusions at the
xpense of data display, and tends to eliminate "less significant"
tems of data that are nevertheless needed for model construction.
e have found that there is no such thing as an insignificant measure-
ent. All the data that are considered reliable should be included
n model construction. While knowledge of the error structure of
he data is helpful, this is often lacking.

A prime requirement for all enzyme studies should be deter-
ination of what is occuring in vivo. This may require more (and
erhaps different) information about an enzyme than is observed in
 typical mechanistic study, especially a suitably broad range of
nitial conditions. For example, studies done at pH or ionic strength
here activity is maximized may have inadvertently placed the enzyme
n an environment completely foreign to that in vivo and may yield
ompletely unphysiological data. It is a common finding that
ffectors other than substrates or products are important controllers
n vivo.

It is preferable to have initial velocity data for an enzyme
n both the forward and reverse directions under the same set of
onditions. All too often the forward and reverse reactions are
un under the different sets of conditions that separately maximize
he rates in the two directions.

It is also possible that enzyme mechanisms in vivo differ from
hat is observed with the isolated enzyme in vitro. Bardsley and
hilds (1975) have shown that complex allosteric reactions can be
ade to appear more simple under the saturating conditions often
sed. Hurst (1974) has found that concentrations needed to identify
igmoid kinetics differ from those that would be sufficient to specify
yperbolic kinetics.

Data for the modelling of individual enzymes usually come from
n vitro experiments with purified enzymes, most commonly initial
elocity studies. These experiments are relatively easy to perform
nd the results may be interpreted using pencil and paper. Observing
apid reaction kinetics, where the enzyme and its intermediates are
n a transient state, requires specialized equipment and yields
bsolute values of rate constants for single steps in a mechanism.

In some favorable cases isotopic tracer experiments may supply useful
information. Simple-minded interpretation of any enzyme data can lea
to errors. A common shortcoming is neglecting ionization and chelati
equilibria so that the concentrations of reactive species are not pro
erly specified. Another is the failure to properly extrapolate back
zero time when observing product accumulation; if one waits a suffici
ently long time for a product accumulation curve to become linear, in
hibitory products may accumulate. Cornish-Bowden (1975) has recently
veloped a method for doing this more accurately that is based on stat
ical considerations.

Besides the types of data available, one must be concerned with
source and conditions under which the data were obtained. Often there
are differences between one laboratory and another for anything more
plex than an isolated enzyme. This includes obvious things such as p
cedural differences, organ, and animal differences. In addition one
must take into consideration much more subtle things such as strain di
ferences, feeding and light schedules relative to the time of death,
yearly and daily periodicities, and climatic conditions. We have en
countered a case where an investigator could not quite reproduce previ
results after moving to a new institution on a different continent.

II. Simple Models

A considerable number of relatively simple models have been con
structed in biochemistry. A brief description will be given of models
involving just a few enzymes, in the expectation that this will assist
in understanding the more complex multi-enzyme models described later.
There now exist many models of single enzymes or of a very small numbe
together; we have reviewed some of these, together with the technique
for building them, elsewhere (Garfinkel et al., 1976b). Two sub-class
of such models merit special attention: those which have been applied
for some useful purpose, and those in which the enzyme is considered i
situ rather than in vitro.

One application of simple enzyme models is in clinical enzymology
where one measures the amount of a given enzyme in a body fluid, usual
serum, or the amount of some organic substance, by a procedure involvi
enzymes. Considerable empirical effort has been devoted to determinir
the optimal conditions under which to perform such assays. Here the e
gineering approach of finding a mathematical function which describes
the assay and then determining the desired reagent concentrations by
formal optimization techniques with that mathematical function appears
useful. These mathematical functions are the rate laws of the approp-

iate enzymes, and methods for determining them are well-established in nzyme kinetics.

We have worked with the serum transaminase assays (Garfinkel, 1973; ondon et al., 1974, 1975), and have been able to determine what expermental conditions will yield a given percentage of the true serum enyme activity. We have shown that as this desired percentage is increased, presumably in the hope of greater reliability, the cost of the etermination above 90 percent or so of maximal activity increases much aster. Since these enzymes have two substrates, each of which could e varied, there are infinitely many pairs of conditions yielding a given ctivity percentage; we have been able to devise a method for picking the neapest one. These procedures can be applied to those enzymes whose echanisms are known.

An outstanding example of enzyme modeling involving just a few enymes in situ is that of Rapoport and co-workers who have been intensvely modeling the mammalian red blood cell. They have been concerned ith the glycolytic pathway, which involves more than a few enzymes, but ave focused their efforts on the most important ones, and have neglected ost of the enzymes in the pathway as not being controlling. They have tudied the relevant isolated enzymes under simulated intracellular (phyiological) conditions, simulated them singly, and combined the results.

Gerber et al. (1974) were able to model red blood cell glycolysis n the basis that hexokinase is the controlling enzyme in regulating lycolysis, phosphofructokinase exerts a secondary effect, and that the ther enzymes may largely be neglected. Although many cells contain ore hexokinase than phosphofructokinase, the former was found to be the imiting enzyme in erythrocytes. They isolated and purified the hexoinase, determined its intial velocity kinetics, fitted a mathematical odel to it, and determined activation and inhibition constants. They oncluded that under physiological conditions the basic glycolytic rate n the red blood cell is adjusted to about 10 percent of the capacity f hexokinase by the concentrations of MgATP, free Mg^{++} (an activator), ,3-diphosphoglycerate and glucose-1,6-diphosphate (both inhibitors), nd that this is modulated by the concentration of glucose-6-phosphate, hich is an inhibitor competitive with MgATP. They also concluded that ree ATP was not controlling.

Kühn et al. (1974a,b) isolated and examined red blood cell phosphoructokinase, concluded that activation by K^+ was mandatory for activity nd that this might be assisted by NH_4^+, especially under pathological onditions. The enzyme was inhibited by ATP, with modulation by other ffectors.

The same group of workers also examined hemoglobin in situ, (Ger ber et al., 1973) since it does affect the environment of the above tw enzymes, and found that it binds about 20 percent of the cellular ATF and approximately 39 or 73 percent of 2,3-diphosglycerate under oxyge ated or deoxygenated conditions respectively. They also found that t concentration of free Mg^{++} went from .7 mM to 1.1 mM on going from ox genated to deoxygenated conditions. They were effectively able to re resent interaction of hemoglobin and the two enzymes and the behavior of each of these in situ.

Most recently Rapoport et al. (1976) have analyzed this system a a whole in terms of control strengths of the five most important enzy involved and steady-state and quasi-steady state behavior of the gly-colytic pathway. In particular, they have defined the mechanisms tha regulate ATP concentration in erythrocytes.

III. Complex Models -- General Considerations

The construction of a complex model in the area of metabolism wa originally a largely intuitive matter. It is steadily becoming more systematized and rigorous logically and mathematically. However, the is still a strong dependence on intuition, ingenuity, and imagination and the learning and incubation periods involved are long. It is dif ficult to report the results of these studies because the resulting models are complex and have properties that are hard to define. We d not yet have general agreement on how to describe models.

Some general guidelines for complex model construction can be su gested:

(1) A thorough knowledge of the relevant biology and of the re-levant literature is a necessity. Mathematical and compute techniques only transform information; they do not create i (In the slightly more pungent generalization which is used the computer industry: "Garbage in, garbage out.").

(2) An important tool in the construction of hypotheses, Occam' razor, here takes the form of the "principle of parsimony": Keep models as simple as possible. In a complex biological situation this may amount to neglecting entities which are known to be present on the ground that they are not importa or not controlling. However, one must distinguish between simplifying a system for the sake of simplicity and actuall showing that certain parts of a system have a negligible ef fect in a given situation.

(3) It is desirable to have a model account for as much relevan

information as possible. This may be considered an alternative form of the principle of parsimony.

(4) An important tactic is to break a complex model down into several simpler parts, study each of these as thoroughly and rig-
orously as possible, and then recombine them.

(5) It is possible to construct mathematically elegant models
which have very little biological utility, because the biology was simplified for the sake of mathematical tractability.
Biological realism is an important requirement.

(6) As a complex model evolves, it tends to become both more sophisticated and more detailed due to the availability of additional experimental information, and improvements of simulation techniques. It is then necessary to draw a line around
what is being modeled, and to approximate inputs from outside
this line. Likewise, it is both possible and necessary to
decide the degree of complexity of what is modeled on the
basis of how much biological information is available. The
evolution of a model with time is discussed further below in
connection with cardiac metabolism.

(7) Simulation is normally a negative process: it disproves theories. Although disproving a theory may be a valuable activity
in its own right, proving one is more difficult. It requires
either disproving all but one possible alternative when there
are known to be a small number, or establishing a hypothesis
which combines information (preferably large amounts of it)
in a new way to explain hitherto unexplained effects and make
predictions that can be verified experimentally.

A general problem is that of communication. Most published experimental work reports on a final workable relationship or procedure, and
does not usually describe the unsuccessful trials. In simulation these
to impart valuable information, and some sort of description is necessary. The usual requirement of scientific description, that a reader
be able to repeat the work from the description, may require more information than can economically be transmitted within the format of a
journal paper, and most readers of that paper will not want to rebuild
model after reading about it. A possible solution to part of this
problem is to put many of the specialized details into an archive or
data depository; this has been discussed at length elsewhere (Gar-
kinkel, 1975). However, the fact that a realistic biological model
contains many elements which are highly interconnected so that many

things influence other things, renders description in journal papers
difficult even with deposition of details.

The problems of communicating a long written account of model co
struction and behavior are not the only ones. It is even difficult t
describe complex models orally given unlimited time. Often one is wo
ing at the limit of the human memory, trying to follow too many thing
or to remember too many obscure scraps of information. Such problems
may arise with even single enzymes. As a result, decisions about whi
sets of data to use in a complex model may become subjective.

Converting from One Organ to Another

One property of a complex metabolic model that is worth consider
here is, how much does a model of a given area of metabolism in one t
sue tell you about the same area of metabolism in another tissue? Co
struction of such a complex model represents a sufficiently large eff
that one should then try to make maximal use of the result. While th
metabolic chart is similar for all tissues, there are considerable qu
itative differences, depending on the detailed metabolism as well as
tissues involved. Glycolysis is more uniform from one tissue to anot
than is the pentose shunt, which may serve completely different functions in different tissues. To a limited extent, a model of one tiss
may suggest the internal milieu of another as a first approximation.
may help place subunits, such as enzymes, in context, and help to def
the problems that must be solved in simulating the behavior of the su
unit in situ. It may suggest which parameters are important. Such
carryover is limited, and a definitive model of a given tissue should
be based on information about that tissue. However, when there is in
sufficient information about the tissue one is modeling, one may be
forced to take information from elsewhere, and the task of modeling t
metabolism of one organ may be considerably simplified by insights ga
from modeling another. At the least, a model of another tissue will
dicate what problems should be looked for; and it may offer qualitati
guidance.

Converting a model from one organ to another is complicated by
the existence of isozymes, which are genetically determined variants
of an enzyme. They all catalyze the same reaction but their kinetic
constants and control properties differ. In some tissues there are
mixtures of isozymes, but in others only one (not always the same one
If the available data apply to the correct isozyme but the wrong anim

vice versa, we must estimate the properties of the desired isozyme.
ometimes the function of a given isozyme distribution is clear. Such
situation exists with the glucose phosphorylating enzymes of liver;
hese phosphorylate blood glucose when its level is high, and permit
he reverse reaction, gluconeogenesis, when blood glucose is low. In
ther cases the reason for these variant forms of the same enzyme is
nclear; in particular there has been considerable discussion of the
easons for the existence of the H and M isozymes of lactate dehydro-
enase. The H isozyme was originally supposed to permit the oxidation
f pyrvuate in aerobic tissues (e.g., heart) but this hypothesis has
een questioned on several grounds (Vessel, 1972). Pyruvate inhibition
f lactate dehydrogenase (H isozyme) is not regulating _in vivo_, and
he heart also contains some of the M-type isozyme. Tissue capacities
ary considerably between organs within the same animal. If data from
he tissue needed are not available it is perhaps best to obtain an
stimate from the same organ of another (preferably similar) animal;
issue capacity tends to fall as body weight rises. In a few cases,
mall groups of enzymes are known to have a constant ratio among them-
elves, "constant proportion groups" (Pette et al., 1962). These
nzymes are evidently coordinately controlled genetically. If one has
ata on one or two members of the group, the values of the others can
e calculated. Attention must be given to variations in tissue capac-
ties due to dietary or other environmental differences.

pecific Considerations Concerning Interconversion of Heart and Liver odels

In the course of our work over the years, we have worked with
odels of both liver and heart metabolism, and have "gone back and
orth" by using a model of the one organ to help in simulation of the
ther. Although various aspects of this work have been published (for
oth organs) much has not, and this is briefly described in the approp-
iate sections below.

In such work one must not lose sight of the basic differences of
he individual organs. The heart principally converts chemical energy,
referably in the form of glucose, lactate or fatty acids, to mechanical
ork. It must respond rapidly to increased workload, so that under
onditions of general basal activity its enzymes are relatively less
ctive and its maximal activities greater than those of the liver.
enerally its metabolism is simpler than that of the liver, especially
ith respect to the Krebs cycle and mitochondrial metabolism.

The liver is believed primarily to be a homeostatic organ; a pri
cipal function is the maintenance of the body's environment. It does
this by interconverting a large number of different substrates in
response to environmental changes. Since certain organs prefer or
require glucose as a substrate, the liver must often convert lactate
(produced by muscle under conditions of virgorous exercise) into gluc
with some aid from the kidneys. This gluconeogenic role of the liver
is one of the major concerns in our recent work.

The different metabolic requirements of the two organs has resul
in significantly different distributions of Krebs cycle enzymes. The
heart must produce ATP rapidly under high load conditions causing
higher cycle fluxes in the heart. While fat synthesis is an importan
liver function, it is not present in heart tissue.

Individual enzyme submodels which were changed to reflect the
properties of the liver isozyme were pyruvate kinase, hexokinase,
phosphofructokinase,aldolase, and lactate dehydrogenase. A number of
enzymes which are not present in heart were added. These include
glucose-6-phosphatase, fructose-1,6-diphosphatase, pyruvate carboxyla
phosphoeonolpyruvate carboxykinase, carbamyl phosphate synthetase, AT
citrate lyase, acetyl-CoA carboxylase, acetoacetyl CoA transferase,
acetoacetyl CoA hydrolase, and β-hydroxybutyrate dehydrogenase. Urea
and fatty acid synthesis were represented as stoichiometric reactions

Magnesium chelation equilibria were added to the mitochondria an
the NAD-linked isocitrate dehydrogenase altered to accommodate the
magnesium-chelated forms of substrate and product.

Simulating mitochondria and cytoplasm together requires distri-
buting appropriate compounds between them, and representing their
interaction. Accordingly, transports for inorganic phosphate, glutam
glutamate-aspartate, malate-inorganic phosphate, malate-citrate, and
α-ketoglutarate-malate have to be included, as well as the Mg^{++} chela
equilibria for those substances that bind it.

While determining mitochondrial and cytosolic redox potentials i
important with both organs, the relative sensitivity to sources of
experimental error differs. The generally low levels of perfusate
lactate in heart experiments makes the amount of extracellular lacta
insignificant. However, in the case of livers perfused with lactate
the interstitial fluid contains significant amounts of lactate, so
that errors in measuring its volume can easily lead to erroneous valu
of tissue lactate, and correspondingly the calculated redox potential
On the other hand, mitochondrial redox potential can be determined
more easily in liver than in heart since the liver excretes significa

antities of acetoacetate and β-hydroxybutyrate, whose ratio depends
 mitochondrial redox potential. Although more data on internal
ncentrations and associated perfusate concentrations of acetoacetate
d β-hydroxybutyrate would perhaps allow us to establish a more rig-
ous transport relationship in liver, we have been able to back-cal-
late a mitochondrial redox potential transient based on transient
rfusate data of these two ketone bodies.

Complex Models -- Specific Examples

mulation of Liver Metabolism

A moderately comprehensive simulation of this area of metabolism
 the perfused rat liver was carried out in a series of papers by
rfinkel (1971a,b), Anderson and Garfinkel (1971), Achs et al. (1971),
d Anderson et al. (1971). We were able to carry out some sophisti-
ted systems anlysis here, both to determine the relative importance
 the several parts of the system, and to see how it operates as a
stem. It is difficult to do such analyses by looking at the isolated
rts of a system or by qualitative examination only.

This work falls into two parts: (1) simulation of the Krebs cycle,
nsisting of relatively crude overall models of the individual react-
ns and a detailed systems analysis, followed by (2) simulation of
e gluconeogenetic pathway with more detailed enzyme models but less
tailed systems analysis. This type of analysis was necessary because
e system was then too complex to compute through economically. Both
udies are based on the experimental work of Williamson et al. (1969a,
c,d). A Krebs cycle model consisting of 35 differential equations
presenting 34 chemical reactions was developed. (It is hoped to
rry out a more detailed analysis with more complete enzyme models
ter). This was matched to the experimental data and then its oper-
ion as a system examined. The relative importance of the enzymes
d to some surprises: NADH-activated isocitrate dehydrogenase, which
s some very complex and elaborate controls was found to exert very
ttle control on the system as a whole; its elaborate controls mostly
em to cancel each other out under the conditions examined. Glutamate
hydrogenase, which is often considered to be near equilibrium and
osely related to the mitochondrial redox potential (so that it is
ten used as an indicator for this) likewise did not appear very
portant under the conditions examined. Conversely, the most important
zyme in this system was found to be pyruvate carboxylase, which
fectively acts as the input of carbon chains to the Krebs cycle and

gluconeogenesis under the conditions examined.

The control properties of some important small molecules were also examined here. The adenine nucleotides were found to be relati unimportant as control factors. To some extent this has been verifie by the subsequent work of Klingeberg (1970) concerning how their mitc chondrial content is affected by oxidative phosphorylation, in a mann quite different from that in the cytoplasm. Their principal effect was as an inhibitor of the enzyme citrate synthase, as suggested by experiments with the isolated enzyme, whereas other controls so sug- gested were unimportant in the complete system. The pyridine nucleo- tides were of moderate importance because of their effect on the amou of oxaloacetate present via the malate dehydrogenase equilibrium; acetyl CoA concentration was of greater importance. By far the most important factor was the concentration of oxaloacetate, which tends to be limiting in the liver. The relative importance of the several system components was assessed by a variety of system analyses, and this component repeatedly turned out to be the most important of all.

By far the most important control of gluconeogenesis appears to be the rate at which substrate is fed into it. In this case, it is mechanism by which pyruvate is converted to oxaloacetate by pyruvate carboxylase and then to phosphenolpyruvate by phosphenolypruvate car- boxykinase. Everything else acts as a secondary modifier. The inter mediates in the pathway do not change concentration much when the overall rate changes considerably. The most important involve the so-called "futile cycles" about fructose-1,6-diphosphatase and phos- phofructokinase, and glucose-6-phosphatase and hexokinase. Both cycl were found to be active and operating at an appreciable rate. In particular, the velocity of fructose-1,6-diphosphatase was fairly constant and the rate of production of fructose-6-phosphate by this couple was actually determined by the rate at which this substance was converted back to fructose-1,6-diphosphate by phosphofructokinase Phosphofructokinase was primarily regulated (inhibited) by citrate concentration, which varies considerably and gets quite high when the is a fatty acid input. The inhibition of phosphofructokinase by citrate, which is quite well-known, is usually described as represent the inhibition of glycolysis by mitochondrial oxidative metabolism. Here this explanation breaks down, as the oxidative metabolism rate varies much less than the citrate concentration, which is primarily a function of the fatty acid input level and the rate of fatty acid synthesis. It is more accurate to claim that here it represents regulation of glycolysis by fatty acid metabolism. Subsequent work,

th experimental (Kemp, 1970) and simulation, indicates that this
fect of citrate is not exerted directly, but as a chelator of Mg^{++}
ich is needed to protect phosphofructokinase from inhibition by free
P. Kühn et al. (1974b) have found that Mg-bound ATP inhibits less
an free ATP or not at all in studies with this same phosphofructokinase
ozyme in the red blood cell. The quantitative extent of the futile
cles appears great enough to account for about 7 percent of the oxi-
tive metabolism of the liver.

A repeated theme that occurred in this work was that the oxidation
f fatty acids, which involves both mitochondrial and cytoplasmic
mponents, as well as transport across the mitochondrial membrane, is
ghly coordinated throughout so that the several members of this
thway are consistently running at closely similar rates.

The overall behavior of the Krebs cycle and related metabolism
re is summed up as follows: "A generalized system property is that
ithin the physiological range nearly all important system parameters
in be changed, smoothly but sensitively and stably, by appropriate
nanges of control parameters" (Garfinkel, 1971b). In particular,
nere were no discontinuities as controlling parameters were varied;
nere saturation effects were observed, they were usually within the
hysiological range, commonly near its upper limit. This is a type
f behavior one would expect from the enzyme kinetics literature, as
aturating concentrations can be measured, and it seems to be believed
nat the physiological range is related to them in this way. One
xception to this pattern is oxaloacetate again: metabolic rates often
urned out to be linear functions of oxaloacetate when they were more
omplex functions of anything else.

Another useful output of this model was to resolve an apparent
xperimental contradiction. It has previously been reported by
illiamson et al. (1969 a,b,c,d) that oleate infusion in perfused
at livers increased the rate of gluconeogenesis, and it had also
een reported by Exton and Park (1967, 1968) that it did not. Needless
o say, their conditions were not exactly identical; in particular
hey were working with different strains of rat. It was possible to
et the model of Williamson's (positive) results to reproduce Exton
nd Park's (negative) results by changing one parameter in the model
y 21 percent, i.e., the outright qualitative difference turned out
o be a small quantitative difference. It is difficult to resolve
uch a disagreement without appropriate quantitative analysis such as
imulation.

Simulation of gluconeogenesis involved the construction of rate

laws for each enzyme in the pathway, with some attention to the contr
of the enzymes. Attention also had to be given to possible compart-
mentation of the substrates and their distribution between mitochondr
and cytoplasm. In particular it was concluded that fructose-1,6-
diphosphate could not all be available to the enzymes metabolizing it
or it would be metabolized much faster, and that about 90 percent of
the reported tissue content was bound, compartmented, or otherwise
unavailable to these enzymes. This may be a consequence of the prese
in liver of two cell types, which may well have differing composition

Another proposed control for this pathway, glyceraldehyde phosph
dehydrogenase, was found not to be particularly important and to exer
only a slight modulating effect. This enzyme appears not to be near
equilibrium under these conditions, contrary to frequent assumption.

In evaluating this work, it should be remembered that it applies
to a specialized situation. Livers from rats which were fasted to
avoid interference from glycogen were perfused with rather high level
of one gluconeogenic substrate and one modulator (fatty acid or ethan
These results may therefore not really be representative of physiological functioning. Likewise the presence of two cell types in live
may limit our understanding until further investigations have been
carried out. One type of cell carries out gluconeogenesis while both
perform glycolysis but probably in different ways. Experiments with
isolated liver cells, which are now available, should particularly
assist in answering such questions.

Simulation of Cardiac Metabolism

Cardiac metabolism is relatively simple in the sense that the
principal duty of the heart is to convert chemical energy to mechanic
work. It has little biosynthetic or secretory activity but is more
complex than other muscular tissues, since it must use a wide variety
of metabolic fuels. Recovery from injury is more difficult, since
the heart can never completely stop working. The heart is also a
strictly aerobic muscle and cannot operate in vivo without oxygen, as
some other muscles do.

The biochemistry by which the heart converts metabolic fuels to
ATP, the chemical form which is the last step before conversion to
mechanical work, has been well studied. Although the amount of available data is relatively large compared to some other organs, there is
still not as much as a model-builder would like. By now there is
beginning to be a considerable amount of data for pathological states,
especially ischemia (insufficient blood flow). There are a number of

animal preparations available which may serve as models; these include
the perfused rat heart with restricted perfusate flow and the dog heart
in situ where flow is interrupted by ligating the coronary artery.

Over the years we have repeatedly examined cardiac metabolism,
starting with supernatant preparations which have glycolytic activity,
progressing to rat hearts perfused under a variety of conditions including anoxia, and then to rat and dog hearts undergoing different
degrees of ischemia. All of these efforts involve considerable attention to individual enzymes, and many of these have been modeled in
considerable detail during the course of this work.

Our first efforts were with supernatant preparations from beef
heart which could be prepared in large quantity and with a high concentration of glycolytic enzymes relative to what is usually obtained
in supernatants. These were investigated at physiological pH (Garfinkel
et al., 1968), and then at a lower pH where the supernatant showed
glycolytic oscillations similar to those observed in yeast (Achs and
Garfinkel, 1968).

In modeling the heart supernatant at physiological pH, it was
possible to achieve a moderately good fit to the experimental data,
although not as good as we usually get nowadays, by postulating additional activations, inhibitions, and enzyme-enzyme interactions over
and above those reported in the literature. Inhibition became cumula-
tively stronger as glycolytic intermediates piled up. An example is
the abnormal behavior of the enzyme triosephosphate isomerase, which
interconverts dihydroxyacetone phosphate (DHAP) and glyceraldehyde
phosphate (GAP). Although its equilibrium ratio, DHAP/GAP = 22,
favors synthesis of dihydroxyacetone phosphate, this enzyme consistently
appears to be moving material in the wrong direction in both supernatants and intact tissues. It makes GAP, in apparent contradiction
to the law of mass action, the discrepancy sometimes being as much as
an order of magnitude. It is both a problem and a virtue of simulation
that such discrepancies cannot be ignored and must somehow be accounted
for or at least represented.

When the temperature is lowered to $15^{\circ}C$ and the pH to 6.25, this
type of supernatant undergoes glycolytic oscillations where the oxida-
tion-reduction state of the cell or preparation rises and falls cyclically. In general such oscillations have been shown to be due to
particular control properties of phosphofructokinase, which is product-
activated and can "turn itself on" (Garfinkel, 1966). In this supernatant preparation such behavior is regulated primarily by the adenine
nucleotides. The curve shapes are not sinusoidal, but there is close

synchrony between the enzymes and the adenine nucleotides; other
substances are in phase with them, and there is a regular periodicity.
However, enzymes other than phosphofructokinase do modulate the oscillation in contrast with the oscillation of the ischemic dog heart
described below.

During this work a more systematic method of constructing such
models began to emerge. Previously our models had been constructed
entirely on an intuitive basis, and the "stiffness" of the differentia
equations was a serious problem. In particular the behavior of enzyme
was calculated using rate laws; this algebraic substitution resulted
in significant savings of computer time. This method has now been
modified so that it can be used to calculate the behavior of separate
portions of a metabolic pathway, and then finally the completed model.
A more complete description of this method is being prepared (Garfinke
et al., 1976a). We then began to examine intact hearts, primarily
perfused rat hearts; these differ from supernatants in having cellular
substructures that carry out significant metabolism. It is therefore
necessary to represent the Krebs cycle and oxidative metabolism of
the mitochondria. Membrane transport into the cell and its mitochondr
now becomes important. Simulation provides information on how a sub-
structure such as a mitochondrion functions in situ and is affected
by its surroundings. We have found that some of the properties of a
rat heart mitochondrion in situ may differ quantitatively from its
behavior in vitro by as much as an order of magnitude (Garfinkel et al
1974). We have also found situations where the behavior of the mitochondrion is quite thoroughly determined by the cytoplasm (other than
the usual situation of increased ADP production). Such interaction
is difficult to study in isolation and is therefore often not taken
into account.

As the heart metabolism model with which we are concerned here is
still evolving -- along with the techniques by which it is being built
-- and no definitive publication of it is presently available to the
reader, a sketchy overview must be provided here. Most of what is now
available is preliminary notes or abstracts (Achs and Garfinkel, 1971,
1972, 1973a,b, 1975; Garfinkel and Achs, 1973, 1974).

This model is concerned with the pathways of metabolism that prov-
the bulk of the heart's energy: glycolysis, the tricarboxylic acid
cycle, fatty acid oxidation (in very limited detail), the major trans-
aminases, etc., together with some of the membrane transport mechanisms
that move substances involved in these pathways in or out of either the
mitochondria or the cell. At the present time there are a total of

68 enzyme or transport mechanisms represented in some detail. As
this area of metabolism does not exist in isolation in the heart,
provision must be made for the kinetic or quantitative effects of the
other parts of metabolism which interact with those included without
also spelling them out in detail. As there are still technical difficulties in running very large models, this is partly an economic
necessity. These are represented by "forcing functions" which help
determine the concentrations or derivatives of chemicals that are
affected by something outside the model.

This model is based on a body of information which is mostly about
the isolated enzymes or transport mechanisms; data about the behavior
of the system as a whole are scarcer. A method of building models in
this situation, based on the enzyme rate laws involved, is described
elsewhere (Garfinkel et al., 1976a). Here unknown concentrations are
chosen to secure the correct behavior of the enzymes with which the
chemical is involved as a substrate or regulator. This method will
not yield a unique model, owing to the typical lack of sufficient experimental data to completely determine it. If there were sufficient
information to uniquely determine a model, it could be built by simpler
and more conventional means. However, the models built in this way
have stood up well to subsequent experimental scrutiny, i.e., predictions
made by this method are likelier to be subsequently proved than disproved.

Thus far four experimental situations have been modeled in detail:
the CO anoxia of Williamson (1966) which is deeper and more thorough
than N_2-based anoxia; the no-substrate to glucose transition of Safer
and Williamson (1973); the work-jump transition of Opie et al. (1971a,
b); and the total ischemia of Krause and Wollenberger (1965) and Wollenberger and Krause (1968). The first three acts of experiments were
done with perfused rat hearts, the fourth, with dog heart in situ.
We are presently examining other ischemic situations as well.

As none of the three sets of experiments with rat heart included
complete measurements of everything needed to determine a model, and
as there was considerable diversity from the one set of measurements
to the other, the data from each experiment was used as an initial
estimate in calculating unknown concentrations for the others. Although
perfused rat heart preparations vary from one laboratory to another,
there is enough simularity to permit this tactic. Unfortunately different ischemic heart preparations differ sufficiently that it has
not yet been feasible to use this tactic with the ischemic heart preparations we are now examining.

Some "highlights" of the findings from these four situations which were not noticed by the experimenters when interpreting their data are

(1) Sudden impostion of CO anoxia (Williamson , 1966) causes a damped
     glycolytic oscillation, which appears to origninate at the level
     of phosphorylase (both a and b) rather than at the level of phos-
     phofructokinase, as in other known instances of glycolytic oscil-
     lation. Waves of material pass down the glycolytic chain but the
     chain is fairly well synchronized in this process, and largely,
     not entirely, under substrate control. At one time point every
     glycolytic enzyme is appreciably displaced from equilibrium, con-
     trary to the common belief that some of them are always at or near
     equilibrium. The "crossover theorem" (Chance et at., 1958), which
     is commonly used to analyze control in the glycolytic chain, fails
     here because it cannot include the phosphorylases which are driving
     the oscillation; their substrate is a solid which is present in
     large quantity and hard to measure accurately. In this particular
     situation this biochemical model permits making a physiological
     prediction: these hearts will stop beating shortly after two
     minutes duration of anoxia.

(2) When glucose was suddenly added to the perfusion fluid for hearts
     which had been exhausted of their glycogen and were burning endog-
     enous fatty acids (Safer and Williamson, 1973), the enzymes in
     the model increased their activity almost entirely under control
     of the increased substrate concentrations. Here phosphofructokin-
     ase, the allosteric enzyme par excellence, displays no allosteric
     control and is instead controlled by its limiting substrate like
     any ordinary enzyme. Here mitochondrial metabolism is controlled
     from the cytoplasm in such a way that the Krebs cycle is "un-
     spanned" (i.e., different parts of it are going at different
     speeds). In this situation the lactate/pyruvate ratio in the
     perfusate, which is sometimes used to calculate the cytoplasmic
     redox potential, does not resemble it and cannot be used to deter-
     mine it.

(3) When lightly working hearts were suddenly switched to working
     very hard by changing the perfusion plumbing (Opie et al., 1971a,
     b), there was a very large increase in the rate of glycolysis,
     enormous relative to the changes in glycolytic substrate levels,
     so that we apparently do not have substrate regulation of the
     enzymes here. An important regulating factor appears to be free
     cytosolic Mg^{++} -- which has to be calculated indirectly. From
     the behavior of the pH-dependent creatine kinase reaction, there
     appears to be a transient drop in pH while the glycolytic rate

is increasing.

(4)   These perfused rat heart models have recently been extended to ischemic heart preparations, especially the preparation of Krause and Wollenberger (1965) and Wollenberger and Krause (1968). This is extreme ischemia, induced by severing the aorta, which causes a glycolytic oscillation. Paterson (1971) observed something similar in rat heart but he obtained less information regarding it. This oscillation is quite different from that observed with the supernatants, since the oscillation is asynchronous and does not even have a fixed periodicity, and the enzymes appear to be considerably desynchronized. It appears not to be based at phosphofructokinase; indeed, it is hard to say what causes it, but the heart gives the appearance of thrashing around in its death agonies. Many unexpected things happen here; e.g., at the later stages phosphohexoseisomerase becomes limiting, even though this enzyme is normally so fast that it can be ignored. Extrapolation of the behavior of the model indicates that the tissue will die in about 4 minutes -- and that it will then contain considerable glycolyzable substrate, primarily glycogen. This agrees with the observation that the heart tissue which has died in cardiac infarcts is often observed to contain stainable glycogen.

We have also simulated less extreme ischemia, primarily based on the rat heart data of Neely et al. (1973, 1975) and Rovetto et al. (1973, 1975). Here, we have been able to account for the hitherto mysterious loss of adenine nucleotides in ischemia. They are converted to adenosine, which acts as a vasodilator, in a massive but futile effort to restore cardiac flow. This can probably be extrapolated to have clinical significance.

An important contrast between the first three conditions and the last one is: in the first three, no matter what is controlling the glycolytic enzymes, they are closely synchronized through some rather vigorous changes in activity, whereas in the fourth they are desynchronized and "thrashing around" as the result of an injury which is immediately lethal. Although more investigation is needed here, there appears to be a strong organized control of these enzymes, which is not easy to break down.

Simulation of Cellular Growth

Perhaps the most comprehensive modeling effort in this area considered is the work of Heinmets (1966) who has been concerned with cellular growth and division and its control by external factors such as hormonal influences. The summary of his work included here is based primarily on his book <u>Analysis of Normal and Abnormal Cell Growth</u>. In his book Heinmets presents a complete mathematical model for cell grow† and then a descriptive model for cell division which is extended to a consideration of cancer which is regarded as primarily an abnormal sta† of cell growth and division.

Growth is defined here as "a process by which mass increases as a function of time". Heinmets' model of growth involves 19 differentia† equations among 25 chemical entites (including 4 genes) undergoing 30 chemical reactions. It represents the RNA and protein sythesis of a generalized cell rather than a specific one. While most of the differential equations are thoroughly nonlinear they do not represent saturation effects or detailed individual mechanisms.

The effects on growth of several factors are studied. These include the transition from dormancy to growth, or adapting to a new culture medium. This is a problem which bacterial cells often face although multicellular organisms do not. Examples are found where the behavior of the model is "counterintuitive", i.e., unpredictable from simple qualitative notions. The general kinetic growth characteristics of various functional elements are different. Although these different elements undergo different time courses in response to sudden medium enrichment, they all double in one generation time. Leakage at the cell membrane and antibiotics which commonly affect such growth-relate† matters as protein synthesis are investigated. Sensitivity analysis (varying parameters and observing the effect on the model) indicates that regulating protein synthesis is a more effective means of controlling growth than nucleic acid synthesis. A cell can live with and recover from a membrane leakage condition lasting several generation times.

Heinmets considers cellular injury and disorganization and how they affect growth, as well as recovery from injury. Biological variability influences resistance to injury. These processes are so inter connected that reliable conclusions cannot be made from a qualitative analysis. Special attention is given to when disorganization becomes fatal, especially for how long a time a given degree of injury or disorganization can be tolerated before it becomes irreversible. The self-organizaing behavior of the system as a means of recovery from injury is investigated. Heinmets is also concerned with critical values of nutrition beyond which a functional system becomes non-operational. When a system is operational it can function over a range of

values, with a minimum value for growth and a possible higher value
beyond which inhibition occurs. There is a definite range of concentrations within which growth can occur, including some non-optimal
conditions. "Simulated experiments" include sudden input changes in
both directions and going from dormancy to growth, where there is
considerable control at the template level. A typical conclusion from
these studies is that "a strong regulatory mechanism is needed in order
for the system to grow smoothly in an environment which varies drastically". This is a common situation for bacterial cells.

Attention is given to therapy, and to which injuries are retriev-
able or are fatal. The conditions under which cellular recovery is
possible are considered: there is a time limit during which recovery
is possible; an otherwise nonfatal injury may become irretrievable if
therapy is delayed. There is some ability to tolerate genetic injury,
if it is not too deep. Cell variability is an important factor in
recovery: "stability variation in the values of individual functional
entities is the determining factor in determining the viability of
the system".

The conclusions drawn by Heinmets are more general than those
described for the other models above. This is partly due to lack of
quantitative data, and even more to the broad area of metabolism which
is being explored. Possibly this type of model is complementary to
the more detailed metabolic models described above in the kinds of
information it provides.

V. Discussion and Conclusions

Of what value are complex models of the type described here, which
are at or near the borderline between biochemistry and physiology?
While they may be expected to generally facilitate interaction between
the two fields, there are a number of specific situations where such
models could prove especially useful.

Since complex models can incorporate more system properties than
the human mind can follow simultaneously, they are useful in those
physiological or pathological situations where the number of significant
biochemical factors exceeds this limit which is, alas, not large. The
worker who obtains a correct interpretation when relying on intuitive
notions of how a given complex system ought to behave is indeed fort-
unate. Where many system properties such as compartment size and ionic
composition inside the cell cannot be measured directly, their estimated
values become more reliable when they have been combined with observed
values to form a plausible mathematical model.

Simulation can also be applied to studying the behavior of an
enzyme as part of a pathway. Enzyme kinetics started out emphasizing
those enzymes which act in isolation (e.g., digestive enzymes such as
trypsin) and the emphasis is still on the isolated individual enzyme,
with relatively little attention to how it interacts with other things
However, most intracellular enzymes do function as part of organized
pathways, and our perhaps limited experience has been that at least
some of these pathways do have closely coordinated control although
biosynthetic pathways where the final product acts as a feedback inhibitor of the first enzyme, may not. This situation involving coordinate
control probably requires such a simulation technique to study thoroug
Furthermore, simulation can be extended to consider modification of
such a pathway by regular physiological processes -- adaptation, biological clock phenomena, and biological variability -- which modify,
but do permit the continued functioning of biological systems.

A possible application of the complex models described here is to
the quantitative manipulation of the biological system represented for
industrial purposes. It is common enough to use unicellular organisms
for chemical transformations and to attempt to control the process in
order to obtain a good yield. Effective use of models might improve
this situation further.

Clinical applications of models like those described here are also
possible. Such a model may be used to assess quantitatively the effec-
of different biological variables on disease conditions and to determine which parameters are worth manipulating in an attempt to cure or
prevent them. One example is a heart suffering from restricted blood
circulation, which will depend more heavily on glycolysis, pile up
lactate, and become more acidic. These techniques might be used to
evaluate the effect of alkalinizing the patient, with its attendant
strain on the kidneys and on other tissues. Another example is invest-
igating the behavior of a cancer which may, as it evolves, switch from
producing one isozyme (usually an adult form) to another (usually a
fetal form). What real difference does this make in its behavior? If
the cancer arose from a virus, with genetic information transmitted by
the reverse transcriptase enzyme, what will be the effect of the appreciable number of errors in transcription that this enzyme makes?

It would appear that effective use of the techniques described
here would greatly facilitate the application of biochemical knowledge
to real biological or pathological situations, where applications of
this knowledge are wanted. It should be cautioned that such applications will probably not materialize immediately.

I. Acknowledgement

This work was supported by grants GM 16501 and HL 15622 from the ational Institutes of Health.

VII. References

Achs, M. J., and Garfinkel, D. (1968) Comput. Biomed. Res. 2, 92.
Achs, M. J., and Garfinkel, D. (1971) Fed. Proc., Fed. Amer. Soc. Exp. Biol. 30, 491 Abs.
Achs, M. J., and Garfinkel, D. (1972) Fed. Proc., Fed. Amer. Soc. Exp. Biol. 31, 349 Abs.
Achs, M. J., and Garfinkel, D. (1973 a) Fed. Proc., Fed. Amer. Soc. Exp. Biol. 32, 344 Abs.
Achs, M. J., and Garfinkel, D. (1973 b) Proc. IFAC Symposium on Dynami and Control in Physiological Systems, Rochester, N. Y.
Achs, M. J., and Garfinkel, D. (1975) Fed. Proc., Fed. Amer. Soc. Exp. Biol. 34, 447 Abs.
Achs, M. J., Anderson, J. H., and Garfinkel (1971) Comput. Biomed. Res 4, 65.
Anderson, J. H., and Garfinkel, D. (1971) Comp. Biomed. Res. 4, 43.
Anderson, J. H., Achs, M. J., and Garfinkel, D. (1971) Comput. Biomed. Res. 4, 107.
Bardsley, W. G., and Childs, R. E. (1975) Biochem. J. 141, 313.
Berman, M. (1965) in Computers in Biomedical Research, vol. II, (Sta R. W., and Waxman, B., eds.), Academic Press, New York, p. 173.
Chance, B., Holmes, W., Higgins, J., and Connelly, C. M. (1958) Nature 182, 1190.
Cornish-Bowden, A. (1975) Biochem. J. 149, 305.
DeLand, E. C. (1967) Chemist - The Rand Chemical Equilibrium Program, Memo RM-5404-PR, Rand Corp., Santa Monica, Calif.
Exton, J. H., and Park, C. R. (1967) J. Biol. Chem. 242, 2622.
Exton, J. H., and Park, C. R. (1968) J. Biol. Chem. 243, 4189.
Garfinkel, D. (1966) J. Biol. Chem. 241, 286.
Garfinkel, D. (1968) Comput. Biomed. Res. 2, 31.
Garfinkel, D. (1971a) Comp. Biomed. Res. 4, 1.
Garfinkel, D. (1971b) Comput. Biomed. Res. 4, 18.
Garfinkel, D. (1973) Human Pathology 4, 79.
Garfinkel, D. (1976) Biosci. Comm. 2, 249.
Garfinkel, D., and Achs, M. J. (1973) Fed. Proc., Fed. Amer. Soc. Exp. Biol. 32, 488 Abs.
Garfinkel, D., and Achs, M. J. (1974) Fed. Proc., Fed. Amer. Soc. Exp. Biol. 33, 373 Abs.
Garfinkel, D., Frenkel, R. A., and Garfinkel, L. (1968) Comput. Biomed Res. 2, 68.
Garfinkel, D., Achs, M. J., and Dzubow, L. (1974) Fed. Proc., Fed. Am. Soc. Exp. Biol. 33, 176.
Garfinkel, D., Achs, M. J., Kohn, M. C., Phifer, J., and Roman, G.-C. (1976a), "A Method of Constructing Metabolic Models Without Repeatedly Solving the Differential Equations Composing Them", in Proceedings of the Summer Computer Simulation Conference, Washington, D. C., p. 493.
Garfinkel, L., Kohn, M. C., and Garfinkel, D. (1976b) CRC Critical Reviews in Bioengineering (in press).
Gear, C. W. (1971) Comm. Ass. Comput. Mach. 14, 176 (with accompanying algorithm, 185).
Gerber, G., Berger, H., Janig, G. R. and Rapoport, S. M. (1973) Eur. J. Biochem. 38, 563.
Gerber, G., Preissler, H., Heinrich, R., and Rapoport, S. M. (1974) Eur. J. Biochem. 45, 39.
Guyton, A. C., Coleman, T. C., and Granger, H. J. (1972) Ann. Rev. Physiol. 34, 13.
Heinmets, F. (1966) "Analysis of Normal and Abnormal Cell Growth", Plenum Press, New York.
Hurst, R. O. (1974) Can. J. Biochem. 52, 1137.

ohnson, L. E. (1974) CRC Critical Reviews in Bioengineering 2, 1.
emp, R. G. (1971) J. Biol. Chem. 246, 245.
ing, J. S. (1975) Clin. Chem. 21, 1349.
lingeberg, M. (1970) Eur. J. Biochem. 13, 247.
rause, E.-G., and Wollenberger, A. (1965) Biochem. Z. 342, 171.
uhn, B., Jacobasch, G., Gerth, C., and Rapoport, S. M. (1974a) Eur.
. Biochem. 43, 437.
uhn, B., Jacobasch, G., Gerth, C., and Rapoport, S. M. (1974b) Eur.
. Biochem. 43, 443.
ondon, J. W., Yarrish, R., Dzubow, L. D., and Garfinkel, D. (1974)
lin. Chem. 20, 1403.
ondon, J. W., Shaw, L. M., Fetterolf, D., and Garfinkel, D. (1975)
lin. Chem. 21, 1939.
eely, J. R., Rovetto, M. J., Whitmer, J. T., and Morgan, H. E. (1973)
m. J. Physiol. 225, 651.
eely, J. R., Whitmer, J. T., and Rovetto, M. J. (1975) Circ. Res. 37,
33.
pie, L. H., Mansford, K. R. L., and Owen, P. (1971 a) Biochem. J. 124,
75.
pie, L. H., Owen, P., and Mansford, K. R. L., (1971b) Cardiov. Res.
uppl. 1, 87.
aterson, R. A. (1971) J. Mol. Cell. Cardiol. 2, 193.
ette, D., Luh, W., and Bücher, Th. (1962) Biochem. Biophys. Res. Comm.
, 419.
apoport, T. A., Heinrich, R., and Rapoport, S. M. (1976) Biochem. J.
54, 449.
augi, G. J., Liang, T., and Blum, J. J. (1975) J. Biol. Chem. 250, 5866
eich, J. G. (1970) FEBS (Fed. Eur. Biochem. Soc.) Lett. 9, 245.
eich, J. G. (1974) Studia Biphysica 42, 165.
oman, G.-C., Garfinkel, D., and Marbach, C. A. (1976) National Com-
uter Conference Proceedings, New York, June 7-10, (in press).
ovetto, M. J., Whitmer, J. T., and Neely, J. R. (1973) Circ. Res. 32,
99
ovetto, M. J., Lamberton, W. F., and Neely, J. R. (1975) Circ. Res. 37,
42
afer, B., and Williamson, J. R. (1973) J. Biol. Chem. 248, 2570.
essel, E. S. (1972) in "Metabolic Inhibitors" vol. 3 (Hoechster, R. M.
and Quastel, J. H., eds) Academic Press, New York, p. 383.
illiamson, J. R. (1966) J. Biol. Chem. 241, 5026.
illiamson, J. R., Browning, E. T., and Scholz, R. (1969 a) J. Biol.
hem. 244, 4607.
illiamson, J. R., Scholz, R., and Browning, E. T. (1969 b) J. Biol.
hem. 244, 4617.
illiamson, J. R., Scholz, R., Browning, E. T., Thurman, R. G., and
Fukami, M. H. (1969 c) J. Biol. Chem. 244, 5044.
illiamson, J. R., Browning, E. T., Thurman, R. G., and Scholz, R.
(1969 d) J. Biol. Chem. 244, 5055.
ollenberger, A., and Krause, E.-G. . (1968) Am. J. Cardiol. 22, 349.

From the multicellular level the discussion now turns to that of single organisms as Professor Riggs demonstrates how mathematical mode of feedback loops can contribute to our understanding of human endeavo including his own backpacking trip last year on the Appalachian Trail.

HOW MODELS OF FEEDBACK SYSTEMS CAN HELP
THE PRACTICAL BIOLOGIST

Douglas S. Riggs, M.D.
Professor of Pharmacology and Therapeutics
State University of New York
at Buffalo
Buffalo, New York, 14057

. Feedback Systems

ntroduction

The fundamental importance of feedback systems -- especially regul-
tory feedback systems -- is universally recognized by engineers and
iologists alike. In order to regulate the temperature of a room, the
ngineer links together certain well-defined components (thermostat,
lectrical relay, oil burner, air-circulating fan) to make a feedback
oop. An important feature of such man-made loops is that they are
ctuated by _error_, defined as the difference between the actual value
f the regulated variable (here room temperature) and the desired or
etpoint value (here the chosen setting of the thermostat). As the room
ools, the error soon becomes large enough to be sensed by the thermostat,
he thermostat signals the furnace to fire, hot air is circulated, and
he room warms up until the error becomes small enough to turn off the
hermostat again. Engineers have developed a powerful array of mathe-
atical techniques which allow them to analyze in detail the time-depend-
nt behaviour of such regulatory systems, and to predict how well the
ystem will perform under a variety of circumstances.
The biologist is also interested in temperature regulation, and
articularly in the ability of such homoiotherms as mammals and birds to
aintain an astonishingly constant body temperature in spite of great
ariations in ambient temperature (See Table 2). But in studying temp-
rature regulation (and other naturally-occurring feedback systems) the
iologist is confronted by a series of formidable difficulties which the
ngineer does not usually have to contend with: 1) The biologist can-
ot choose well-defined, man-made components for his feedback loops; he
ust accept whatever regulatory mechanisms have been selected as biologically best during millions of years of evolutionary adaptation. These
'best" choices are not necessarily the easiest ones to describe mathemat-

ically! 2) True setpoints and error signals rarely exist <u>within</u> the
moist and mobile world of living systems, though <u>external</u> setpoints o:
play a key role in regulating the orientation of an animal with respec
to some external target. Conventional wisdom to the contrary notwithstanding, there is no thermostat in your brain, nor any need for one.
3) Biological feedback systems often comprise many different feedbacl
loops interlinked in complicated ways. Regulation of body temperature
provides an outstanding example. 4) As in the case of certain engine
ing systems, biological feedback systems may contain positive loops wh
tend to augment the effect of an external disturbance. 5) Engineers
can often choose to work with linear components, i.e., components whos
output ("response") is directly proportional to the input ("stimulus")
at least over a considerable operating range. In contrast, many components of biological feedback loops are highly nonlinear. This feat
complicates the mathematical analysis. 6) All too frequently, the bi
logist cannot even identify, -- let alone measure --, some of the impc
ant variables in the feedback loops he wants to study. These obstacle
have discouraged attempts to characterize quantitatively the overall
behavior of biological feedback mechanisms.

In this paper I shall describe and illustrate a comparatively sin
technique of mathematical analysis which may allow the biologist to
measure the effectiveness of a biological feedback system in spite of
the difficulties described above. As presented here, <u>the technique is</u>
<u>useful only for studying the steady-state behaviour of feedback loops</u>
<u>in response to constant inputs</u>. Much more complicated mathematical
methods have to be used to characterize the transient time-dependent
behavior of a biological feedback system as it moves from one steady
state to another, or when it exhibits some kind of oscillatory or limi
cycle behavior.

Causal Relationships. Symbol-and-Arrow Diagrams.

A feedback loop is a closed unidirectional cycle of causal relati
ships. In the steady state, causal relationships can best be depicted
by a symbol-and-arrow diagram. We shall use

$$A \longrightarrow B \qquad\qquad (1)$$

to show that an increase in A causes an increase in B; a decrease in A
causes a decrease in B. We shall use

$$A \dashrightarrow B \qquad\qquad (2)$$

o show that an increase in A causes a decrease in B; a decrease in A
auses an increase in B.   Note that relationships (1) and (2) are <u>uni-</u>
<u>irectional</u> <u>relationships</u>; if·a primary change were made in B, nothing
ould happen to A.   We shall use

$$A \longleftrightarrow B \tag{3}$$

r

$$A \longleftarrow\text{-->} B \tag{4}$$

o show that A and B are in equilibrium with each other.   For example,
n relationship (4) a primary increase in A causes a secondary decrease
n B <u>and</u> a primary increase in B causes a secondary decrease in A.   A
ingle two-headed arrow is used to depict an equilibrium between A and
because an equilibrium is a <u>single</u> reversible relationship described
y a <u>single</u> equation.   Finally we shall use

$$A \, B \tag{5}$$

r

$$A \, B \tag{6}$$

o show that A and B are members of a <u>negative feedback</u> loop, and

$$A \, B \tag{7}$$

r

$$A \, B \tag{8}$$

to show that A and B are members of a <u>positive feedback</u> loop.   Note that
when two variables have a feedback relationship with each other, two
<u>different</u> unidirectional arrows must be used, because the change in B
caused by a change in A must be described by a first equation, while
the change in A caused by a change in B must be described by an entirely
<u>different</u> second equation.

Negative feedback loops act to decrease the effect of an external disturbance on the variables in the loop. For example, suppose that

$$I \dashrightarrow L_1 \quad L_2 \tag{9}$$

depicts a feedback loop composed of the loop variables L_1 and L_2. L_1 can be disturbed by a change in the independent input variable, I. Diagram (9) shows that an increase in I will cause a primary decrease in This decrease in L_1 will cause a decrease in L_2 as indicated by the unbroken arrow in the loop. But by the second relationship indicated by the broken arrow in the loop, the decrease in L_2 will cause a secondary increase in L_1 which tends to oppose the primary change. A precisely similar argument will show that (6) is also a negative feedback loop.

Positive feedback loops act to increase the effect of an external disturbance on the variables in the loop. Using the same reasoning as before, it is easy to see that in the system of diagram (7) or of diagram (8) any primary change in one of the loop variables will tend to be enhanced by the operation of the feedback loop.

Feedback loops often consist of more than just two variables and the appropriate intervening arrows. With multivariable loops, it is useful to know that a loop containing an odd number of broken arrows is a negative feedback loop; a loop containing an even number of broke arrows is a positive feedback loop.

Magnification: A Measure of the Effectiveness of Feedback Loops in the Steady State.

In order to compare the effectiveness of different feedback syste or to study variations in the effectiveness of a single feedback syste under different conditions, we need to have some measure of how effectively a negative feedback loop decreases (or how effectively a positive feedback loop increases) the change in a loop variable caused by an external disturbance (input). As a general index of the effectiveness of feedback in the steady state, we define [Mag], the magnification due to feedback:

$$[\text{Mag}] \equiv \frac{\left. \dfrac{\partial L_i}{\partial I_p} \right|_{\text{closed}}}{\left. \dfrac{\partial L_i}{\partial I_p} \right|_{\text{opened}}} \tag{10}$$

where L_i is any particular variable in the feedback loop,

I_p is any particular input variable which can influence L_i,

the numerator is the change in L_i for a given change in I_p when the loop is <u>closed</u> so that feedback occurs, and

the denominator is the change in L_i for a given change in I_p when the loop has been interrupted (<u>opened</u>) so that feedback can no longer occur.

Magnification is a finite number equal to or greater than zero. For negative feedback it is less than unity; for totally ineffective feedback it is unity; for positive feedback it is greater than unity. In the vast majority of nonlinear systems, [Mag] differs from one steady-state operating point to another, so that <u>the open-loop change and the closed-loop change of (10) must both be evaluated at the same operating point</u>. (For a definition of "operating point", see the next section.)

If the loop has to be interrupted surgically or pharmacologically in order to study the open-loop situation, it may be difficult or impossible to re-establish the original closed-loop operating point. Fortunately [Mag] can often be estimated indirectly without actually opening the loop in such a heavy-handed way. Let L_i be any particular loop variable which can be measured and which can also be altered experimentally by manipulating some appropriate input, I_p. Let L_j be a second loop variable which can be measured and which can also be altered experimentally by manipulating an appropriate input, I_q. In a first series of experiments, we make primary changes in L_i by changing I_p stepwise, waiting long enough at each step for the system to approximate a new steady state. We plot the observed steady-state values of L_j against the observed steady-state values of L_i and fit an appropriate curve to the points. The equation for this curve is a specific version of the general equation for L_j as a function, f, of L_i

$$L_j = f(L_i, I_q) \tag{11}$$

In a second series of similar experiments we make primary changes in L_j by manipulating I_q so as to obtain a specific version of the general equation for L_i as a function, g, of L_j

$$L_i = g(L_j, I_p) \tag{12}$$

Now it can be shown (Riggs, 1970) that

$$[Mag] = \frac{1}{1 - \frac{\partial f}{\partial L_i} \frac{\partial g}{\partial L_j}}$$

(13)

where the first derivatives $\frac{\partial f}{\partial L_i}$ and $\frac{\partial g}{\partial L_j}$ are to be evaluated at the oper ating point values of L_i and L_j obtained by solving equations (11) and (12) simultaneously. (The absolute value of the product $\frac{\partial f}{\partial L_i} \cdot \frac{\partial g}{\partial L_j}$ is [O the steady-state open loop gain of the system. Unlike [Mag], [OLG] ca also be defined for transient states and is therefore quite rightly pr ferred by engineers as a measure of the effectiveness of feedback. Ho ever, since we are confining our attention to steady-state analysis, i is best to use [Mag], which has an easily-understood everyday meaning, rather than [OLG] which does not.)

In the preceding discussion of [Mag] and its measurement, we con fined attention to single feedback loops. But [Mag] can also be used to measure the effectiveness of feedback in those complicated multiple loop systems which are so often encountered in biology. Much of the re of this paper will be devoted to a detailed analysis of just such a system.

Model Systems, Steady-state Operating Points, and Symbol-and-Arrow Diagrams

Real biological systems are composed of such a rich and bewilderir array of variables that the practical biologist must always begin by re ducing his real system to a manageably simple model. A steady-state model will consist of m independent input variables, I, n dependent var iables, D, and the n working equations (one for each D) which specify how each of the dependent variables depends (in a causal sense) upon th other variables of the system (Riggs, 1970, Sec. 1.5). The working eq uations may be theoretical, or empirical, or a mixture of both types.

In any such model system, we are free to choose an arbitrary value for each of the m input variables. The resulting set of m values, to gether with the n working equations, will uniquely determine the values of all of the n dependent variables, provided that the values chosen fo the inputs are compatible with a steady state. We define the corres ponding operating point (a point in (n + m) - space) as the vector, $[I_1, I_2, \ldots, I_m, D_1, D_2, \ldots, D_n]$, consisting of the steady-state values o inputs and dependent variables.

In a symbol-and-arrow diagram for any such model system, the inde-

pendent inputs are the variables towards which no arrows point. Any
variable which has one or more arrows pointing towards it is a dependent
variable which may or may not also be a member of a feedback loop.

Linearization Around an Operating Point

As was pointed out previously, the first derivatives, $\frac{\partial f}{\partial L_i}$ and $\frac{\partial g}{\partial L_j}$ in
equation (10) must both be evaluated at the <u>same</u> operating point because
in nonlinear systems the magnification due to feedback differs from one
operating point to another. In essence, what we do is to study ΔL_i,
the <u>small</u> deviation of L_i from its operating-point value which occurs
in response to ΔI_p, a <u>small</u> deviation of an appropriate input variable
from its operating point value. But if we are confining our attention
to <u>small</u> deviations from a particular operating point, we can replace
our nonlinear set of n equations by a corresponding set of n linear equations obtained by expanding each equation around the operating point
as a multivariable Taylor's series in which we disregard all terms in
second- or higher-order differences retaining only the linear terms (See
Appendix B). In essence, we thus replace a set of curved hypersurfaces
which intersect at the operating point in (n + m) - space by the corresponding uncurved hyperplanes which are tangent to the curved hypersurfaces at their point of intersection. We can then treat our nonlinear
system as if it were a simple linear system, <u>provided</u> that we confine our
attention to sufficiently small changes in the neighborhood of the chosen
operating point. The actual steps used for linearizing around an operating point are not difficult, and are illustrated in Appendix B.

II. A Concrete Example: Backpacking as a Positive Feedback System.

In backpacking along a wilderness trail between Supply Point I and
Supply Point II one wants to traverse the intervening terrain with a
maximum of comfort and a minimum of work. It must have occurred to other
backpackers as forcibly as it did to me that one is always hampered by
certain incorrigibly positive feedback loops: the more you carry, the
harder you work; the harder you work, the more you eat; the more you
eat, the more food you have to carry; the more food you carry, the harder
you work. Likewise, the more you carry, the slower you go; the slower
you go, the more time you spend between Point I and Point II; the more
time you spend, the more you eat; the more you eat, the more food you
have to carry; the more food you carry, the slower you go.

Let us analyze this multiple-loop, positive, feedback system by using the general methods outlined above.

Symbols, Definitions, and Exemplary Values.

The values listed in Table 1 for "Inputs or Parameters" are rough. correct for the author of this paper while he was backpacking on the A palachian Trail in Northern New England. The values of the dependent variables calculated from these inputs and parameters are given to thre significant figures.

A _parameter_ is an independent variable which is held constant. B in the steady state, inputs must also be held constant. In steady-sta analysis, therefore, the choice of which independent variables to rega as inputs is entirely arbitrary.

Derivation of the Building-block Equations

The most difficult part of the entire analysis is the constructio of a model which is simple enough so that it can be analyzed mathemati cally, yet realistic enough so that the results of the analysis can be used to predict the behavior of the actual biological system.

Reduction of a real system to a suitable model is accomplished largely by making appropriate assumptions. Most, if not all, of these simplifying assumptions can be expressed as equations. Other "buildin block" equations are true by definition, or describe known physical or chemical properties of the system.

Equations Derived from Assumptions:

Assume that the work of climbing is determined entirely by the total vertical ascent, and is independent of horizontal progress Then

$$E_Y = 2.343 \, \overline{W}_{tot} Y/f \tag{14}$$

where 2.343 is Kcal per kg·km.

Assume that the rate of vertical ascent is always at the maximum allowed by p_{max}, the maximum above-basal power consumption.

$$T_Y = E_Y/p_{max} \tag{15}$$

Table 1

Symbols, Definitions, and Exemplary Values

Symbol	Inputs or Parameters	Operating Point Values
W_{body}	weight of body (nude)	60 kg
W_{equip}	weight of clothes and equipment	20 kg
v_{max}	velocity of walking with no load	5 km/hr
k	decrease in v per kg of load	1/12 km/hr·kg
T_{day}	time on trail each day	7 hrs
P_{bas}	basal power consumption	90 Kcal/hr
P_{max}	maximum steady state power consumption <u>above</u> basal	600 Kcal/hr
c	energy supplied per unit weight of food	3750 Kcal/kg
f	fractional efficiency of performing work of ascent	0.15
K	energy used in transporting 1 kg for 1 km	0.8 Kcal/kg·km
X	horizontal distance by trail between supply points	125 km
Y	total vertical ascent between supply points	6 km
	Dependent loop variables	
$W_{f.max}$	weight of food just after resupplying	8.92 kg
$\overline{W}_f$	mean weight of food carried	4.46 kg
$\overline{W}_{load}$	mean weight of non-body load	24.5 kg
$\overline{W}_{tot}$	mean total weight	84.5 kg
E_Y	total energy expense of ascending vertically	7920 Kcal
E_X	total energy expense of walking horizontally	8450 Kcal
E_{bas}	total energy expense of basal activities	17,100 Kcal
E_{tot}	total energy expended between supply points	33,500 Kcal
T_Y	total time used in ascending vertically	13.2 hr
T_X	total time used in walking horizontally	42.2 hr
T_{tot}	total time on trail	55.4 hr
D	number of days between supply points	7.91 day
v	velocity of walking horizontally	2.96 km/hr

Assume that the energy required for horizontal transport is direc[t] proportional to the product of total weight and total horizonta[l] distance. (Although no external work is performed in walking on[m] a level path, energy is expended. See, for example, Tucker, 1975).

$$E_X = K\overline{W}_{tot}X \tag{16}$$

Assume that the velocity of horizontal hiking decreases linearly with increasing load from a maximum at zero load to zero at max[imum] imum load, here taken to be equal to body weight.

$$v = v_{max} - k\overline{W}_{load} \tag{17}$$

Assume that the total weight of food obtained in resupplying yiel[ds] just enough calories to equal the total energy expended between supply points, so that the body weight stays constant.

$$W_{f.max} = E_{tot}/c \tag{18}$$

Assume that between supply points food is consumed at a uniform rate so that the mean weight of food is half the maximum weight just after resupplying (See Appendix C).

$$\overline{W}_f = 0.5W_{f.max} \tag{19}$$

Equations Which Are True by Definition:

$$\overline{W}_{tot} = \overline{W}_{load} + W_{body} \tag{20}$$

$$\overline{W}_{load} = W_{equip} + \overline{W}_f \tag{21}$$

$$E_{tot} = E_X + E_Y + E_{bas} \tag{22}$$

$$E_{bas} = 24p_{bas}D \tag{23}$$

$$D = T_{tot}/T_{day} \tag{24}$$

$$T_{tot} = T_X + T_Y \tag{25}$$

$$T_X = X/v \tag{26}$$

Construction of a Symbol-and-Arrow Diagram

Each of the "building-block" equations (14) through (26) has been solved for its causally-dependent variable. I call any such equation a "working equation" because it shows how the corresponding part of the model system works in a cause-and-effect sense.

Putting the building-block equations in working-equation form makes it easy to draw a symbol-and-arrow diagram of the system. In Figure 1, the number near the symbol for a dependent variable refers to the "building-block" equation for that variable. Symbols lacking numbers denote independent variables (inputs or parameters).

Figure 1. A Symbol-and-Arrow Diagram Showing the Causal Relationships Among the Variables of Equations (14) through (26).

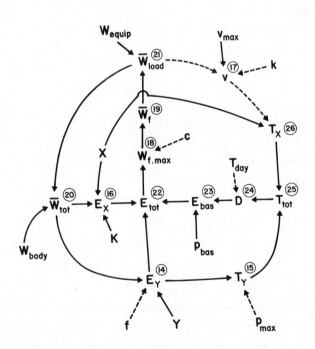

The number near each dependent variable is the number of the corresponding "building-block" equation. Symbols lacking numbers denote independent variables (inputs or parameters).

Scrutiny of Figure 1 shows that there are four feedback loops; ea
contains an even number (here zero or two) of broken arrows. All of
feedback loops in this system are therefore positive.

Combining Equations while Preserving Loops

Considerable simplification can be achieved by combining some of
the building-block equations. The system can then be described by fou
equations. Any further reduction in the number of equations would con
ceal one or more of the feedback loops.

Combine equations (16), (18), (19), and (21)-(24):

$$\overline{W}_{load} = W_{equip} + \frac{1}{2c}\left[KX\overline{W}_{tot} + E_Y + \left[\frac{24p_{bas}}{T_{day}}\right] T_{tot}\right] \tag{27}$$

Combine equations (15), (17), (25), and (26):

$$T_{tot} = \frac{X}{v_{max} - k\overline{W}_{load}} + \frac{E_Y}{p_{max}} \tag{28}$$

Equations (14), (20), (27), and (28) correspond to the symbol-and-arro
diagram of Figure 2 which, for simplicity, shows only loop variables.

<u>Figure 2.   A Much-Simplified Symbol-and-Arrow Diagram.</u>

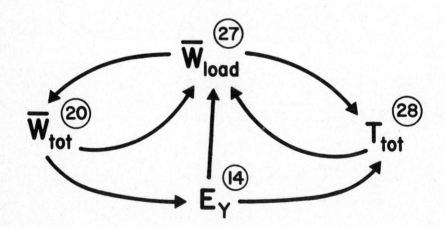

This diagram shows the causal relationships among the four remain
loop variables of equations (14), (20), (27) and (28). The independen
input variables have been omitted.

Finding the Operating Point Values of the Dependent Variables.

To find general algebraic expressions for each dependent variable
in terms of the independent variables would require simultaneous solutic
of equations (14) through (26). Though this would be technically possible, the resulting expressions would be too cumbersome for practical
use. To simplify matters, let us retain X, Y, c, and W_{equip} as the most
interesting inputs. Let the eight other independent variables have the
exemplary numerical values in Table 1. Then equations (27), (28), (14),
and (20) become:

$$\overline{W}_{load} = W_{equip} + \frac{1}{2c}\left[0.8X\overline{W}_{tot} + E_Y + 308.57T_{tot}\right] \tag{27a}$$

$$T_{tot} = \frac{X}{5 - \frac{1}{12}\overline{W}_{load}} + \frac{E_Y}{600} \tag{28a}$$

$$E_Y = 15.62Y\overline{W}_{tot} \tag{14a}$$

$$\overline{W}_{tot} = 60 + \overline{W}_{load} \tag{20a}$$

Solving this set of four simultaneous equations for $\overline{W}_{load}$, we obtain

$$\overline{W}_{load} = 0.5\left[B - \sqrt{B^2 - 4J}\right] \tag{29}$$

where

$$B = 60 + \frac{b}{\alpha} \tag{30}$$

$$J = 60\frac{b}{\alpha} + \frac{a}{\alpha} \tag{31}$$

and

$$a = 3702.86X \tag{32}$$

$$\alpha = 2c - 0.8X - 23.653Y \tag{33}$$

$$b = 2cW_{equip} + 48X + 1419.19Y \tag{34}$$

We can now substitute into equations (32)-(34) whatever values we choo
for X, Y, c, and $\overline{W}_{equip}$, and solve (29) for the corresponding numerica
value of $\overline{W}_{load}$. Having a numerical value for $\overline{W}_{load}$ enables us to find
congruent numerical values for all of the other dependent variables.
These numerical values, together with the values of the independent va
iables, constitute the operating point values of the system. An examp
of a set of operating point values is given in Table 1.

Linearizing Around the Operating Point.

The general philosophy of using Taylor's series to linearize a se
of equations in several variables around a chosen operating point was
explained in a previous section. (For method, see Appendix B.) To ap
this method to more complex models, each equation must be expanded in
terms of partial derivatives, one for each of the variables on the rig
hand side of the equation. This procedure is illustrated in Appendix

Applying this technique of linearization to our backpacking model
(equations (27a), (28a), (14a), and (20a)), we obtain

$$\Delta \overline{W}_{load} = \beta_1 \Delta W_{equip} + \beta_2 \Delta c + \beta_3 \Delta X + \beta_4 \Delta \overline{W}_{tot}$$
$$+ \beta_5 \Delta E_Y + \beta_6 \Delta T_{tot} \tag{35}$$

$$\Delta T_{tot} = \beta_7 \Delta X + \beta_8 \Delta \overline{W}_{load} + \beta_9 \Delta E_Y \tag{36}$$

$$\Delta E_Y = \beta_{10} \Delta Y + \beta_{11} \Delta \overline{W}_{tot} \tag{37}$$

$$\Delta \overline{W}_{tot} = \beta_{12} \Delta \overline{W}_{load} \tag{38}$$

In this set of equations, the β's are the appropriate partial derivatives:

$$\beta_1 = 1$$

$$\beta_2 = -\frac{1}{2c^2} \left[0.8 X \overline{W}_{tot} + E_Y + 308.57 T_{tot} \right]$$

$$\beta_3 = 0.8 \overline{W}_{tot} / 2c$$

$$\beta_4 = 0.8X/2c$$

$$\beta_5 = 1/2c$$

$$\beta_6 = 308.57/2c$$

$$\beta_7 = \cfrac{1}{5 - \frac{1}{12}\,\overline{W}_{load}}$$

$$\beta_8 = \cfrac{X/12}{\left[5 - \frac{1}{12}\,\overline{W}_{load}\right]^2}$$

(39)

$$\beta_9 = 1/600$$

$$\beta_{10} = 15.62\overline{W}_{tot}$$

$$\beta_{11} = 15.62Y$$

$$\beta_{12} = 1$$

Calculating the Magnification Due to Feedback

Suppose that we want to find out how much the entire set of four feedback loops, acting together, magnifies the effect of an increase in X, the distance between supply points, on $\overline{W}_{load}$, the mean load. We keep all other inputs unchanged, simply by letting $\Delta W_{equip} = \Delta c = \Delta Y = 0$. We then solve equations (35)-(38) simultaneously for the closed-loop ratio, $\Delta\overline{W}_{load}/\Delta X\,\big|_{closed}$

$$\frac{\Delta\overline{W}_{load}}{\Delta X}\bigg|_{closed} = \frac{\beta_3 + \beta_6\beta_7}{1 - \left[\beta_4 + \beta_5\beta_{11} + \beta_6\beta_8 + \beta_6\beta_9\beta_{11}\right]}$$

(40)

To find the corresponding open-loop change, $\Delta\overline{W}_{load}/\Delta X\,\big|_{open}$, we interrupt feedback via each loop by setting appropriate β's equal to zero. In doing this, however, we must be careful to preserve all of the β's which are involved in the non-feedback influence of ΔX, our chosen input, on $\Delta\overline{W}_{load}$ both directly (equation (35)), and via ΔT_{tot} (equation (36)).

We must keep these open-loop pathways intact. Let us therefore set β_4, β_5, β_8 and either β_9 or β_{11} (or both) equal to zero, and let us solve the resulting open-loop equations,

$$\Delta \overline{W}_{load} = \beta_3 \Delta X + \beta_6 \Delta T \tag{41}$$

$$\Delta T = \beta_7 \Delta X, \tag{42}$$

simultaneously for the open-loop ratio

$$\left. \frac{\Delta \overline{W}_{load}}{\Delta X} \right|_{open} = \beta_3 + \beta_6 \beta_7, \tag{43}$$

Equation (43) can more easily be obtained by setting $\beta_4 = \beta_5 = \beta_8 = \beta_9$ ($= \beta_{11}$) $= 0$ in equation (42).

Now equations (40) and (43) are written in terms of finite changes. But we are interested in calculating the magnification <u>at</u> a particular operating point. We can therefore allow the finite Δ's to become arbitrarily small, thus obtaining [Mag] directly as the ratio of the closed loop change of (40) to the open-loop change of (43):

$$\frac{\left. \dfrac{\Delta \overline{W}_{load}}{\Delta X} \right|_{closed}}{\left. \dfrac{\Delta \overline{W}_{load}}{\Delta X} \right|_{open}} = \frac{\left. \dfrac{d \overline{W}_{load}}{dX} \right|_{closed}}{\left. \dfrac{d \overline{W}_{load}}{dX} \right|_{open}} = [Mag] \tag{44}$$

$$= \frac{1}{1 - [\beta_4 + \beta_5 \beta_{11} + \beta_6 \beta_8 + \beta_6 \beta_9 \beta_{11}]}$$

Note that the β's must be evaluated by substituting into equations (39) the values which the variables have <u>at</u> <u>the</u> <u>particular</u> <u>operating</u> <u>point</u> <u>of</u> <u>interest</u>. Each β thus assumes a specific numerical value which can be substituted into (44) to obtain the magnification at that particular point. For the operating point values in Table 1, $\beta_4 = 100/750$, $\beta_5 = 1/7500$, $\beta_6 = 308.57/7500$, $\beta_8 = [125/12] \div \left[5 - \frac{1}{12} [24.46]\right]^2$, $\beta_9 = 1/600$, and $\beta_{11} = 6[15.62]$. Substituting these values into (44), we find [Mag] $= 1.088$. In other words, due to the joint operation of the positive feedback loops the increment in load produced by a small increment in distance is about 9% greater than it would have been without any feedback.

How Magnification is Influenced by Changing Various Inputs

Figure 3 shows how [Mag] changes when $\overline{W}_{equip}$ is altered in either direction, when c is decreased, or when both X and Y are simultaneously increased by the same factor so as to keep the "ruggedness" of the trail constant. (I here take the ratio, Y/X, as a measure of "ruggedness"). For ease in making comparisons, the numbers on the horizontal scale show each input as a fraction or multiple of its value at the original operating point of Table 1. Both scales are logarithmic. The following features of Figure 3 are noteworthy.

(1) Decreases in c or increases in W_{equip}, or increases in both X and Y cause increases in [Mag] as would be expected. For all three curves, the increases are at first rather slight, but the curves steepen with frightening rapidity as a critical value* for the particular input is approached. At this critical value, [Mag] becomes infinite so that a stable steady state can no longer exist. This means that (under the assumptions here being made) the unfortunate backpacker would never reach Supply Point II! So abrupt is the final rise in [Mag], that a seemingly modest magnification of, say, 1.5 would, in practice, represent a potentially dangerous situation where unanticipated difficulties might well tip the balance towards disaster.

(2) Changing both X and Y by a factor of F causes the same change in [Mag] as does changing c by a factor of 1/F. In fact, if c, X and Y are all multiplied by the same factor, [Mag] remains constant. The reason is simply that doubling c, the caloric value of the food carried, doubles the energy supplied by a given weight of food. Hence the energy expended in walking and climbing can also be doubled, leaving the magnification unchanged at its original value.

(3) In the present example, increasing the weight of clothes and other equipment by a given factor causes a greater rise in [Mag] than does increasing both X and Y by the same factor. Furthermore, the critical factor for W_{equip} is about 2, while for simultaneous increases in both X and Y it is about 10/3. However, the critical factor for W_{equip} depends very heavily upon what the value of W_{equip} is at the original operating point; obviously the smaller the original value, the larger the factor by which it can be increased before it reaches the critical value.

---

*In this system, the critical value for a given input is the value at which $B^2 = 4J$ in equation (29).

Figure 3. The Magnification Due to Feedback at Various Operating Poir

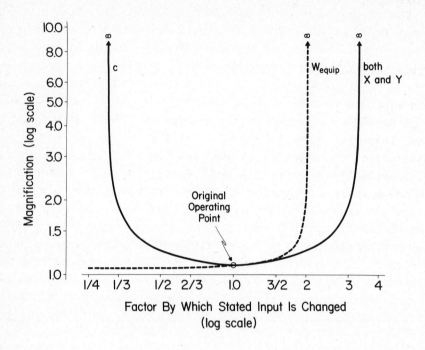

The horizontal scale shows the factor by which the designated inp
(or inputs) differs from its original value at the operating point of
Table 1. The solid line to the left of the original point (open circl
represents decreases in c only. The solid line to the right represent
increases in X and Y only, both being changed by the same factor. The
broken line represents both decreases and increases in W_{equip} only.
Note that the scales are logarithmic.

In backpacking, the importance of minimizing the weight of clothe
and other equipment has been emphasized by the experts time and again.
Not so universally noticed is the fact that obesity is a dreadful en-
cumbrance. Under the assumption that body weight is to remain constan
excess fat must be regarded as equivalent to additional "equipment",
subject to precisely the same penalties as carrying a large camera, or
a needlessly heavy tent. (By the same token, of course, backpacking i
a wonderful way to lose excess weight; many of the people I met on tra
were taking advantage of this!)

In backpacking, maximizing the Kcal/kg of food carried is crucial
If you tried to backpack on an ordinary diet of fresh (or canned) brea

meat, dairy products, fruits and vegetables you would not get very far.
Concentrated foods are essential to raise the value of c = Kcal/kg of
food. The ready availability of dehydrated foods and featherweight
equipment fully accounts for the present popularity of backpacking.

Generalizing the Meaning of Magnification.

In equation (44) we chose to define [Mag] in terms of the change
in $\overline{W}_{load}$ caused by a change in X. Because $\overline{W}_{load}$ is a member of all four
feedback loops, the same function of the β's as the one in (44) would
be obtained for the magnification of the effect of any other input on
$\overline{W}_{load}$, provided that in the closed-loop situation all loops are closed
and in the open-loop situation all loops are open. But these restrictions are not inherent in the definition of [Mag]. Suppose, for example,
that we wanted to find out how much the change in T_{tot} caused by a given
change in Y is magnified by the operation of the two feedback loops
containing T_{tot}, the other two loops remaining closed. With all four
loops closed, we obtain

$$\frac{\Delta T_{tot}}{\Delta Y}\bigg|_{closed} = \frac{\beta_{10}[\beta_5\beta_8 + \beta_9 - \beta_4\beta_9]}{1 - [\beta_4 + \beta_5\beta_{11} + \beta_6\beta_8 + \beta_6\beta_9\beta_{11}]} . \qquad (45)$$

After opening only those loops containing ΔT_{tot} (by letting $\beta_6 = 0$) we
obtain

$$\frac{\Delta T_{tot}}{\Delta Y}\bigg|_{\substack{T_{tot}\ loops \\ open}} = \frac{\beta_{10}[\beta_5\beta_8 + \beta_9 - \beta_4\beta_9]}{1 - [\beta_4 + \beta_5\beta_{11}]} \qquad (46)$$

Hence, for this situation,

$$[Mag] = \frac{1 - [\beta_4 + \beta_5\beta_{11}]}{1 - [\beta_4 + \beta_5\beta_{11} + \beta_6\beta_8 + \beta_6\beta_9\beta_{11}]} \qquad (47)$$

which is a different function of the β's than the one in (44).

This example should make it clear that, in multiple-loop systems,
[Mag] has a very general meaning which can be particularized so as to
let us answer whatever specific questions about the system we care to
pose. A block diagram, such as the one in Figure 4 for the linearized
version of the present system, helps us to see which β's can be set to

Figure 4, Block Diagram for the Linearized System

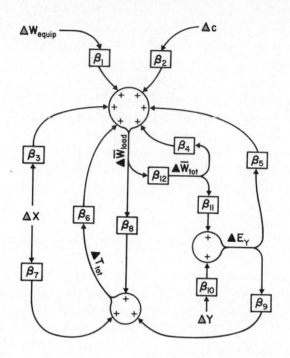

This diagram is a representation of equations (35) through (38). Each block in a block diagram contains a _transfer function_, which, in general, is a function of the Laplace transform variable, s, a complex number. But _in the steady state_, s vanishes, so that (for any given operating point) each transfer function is a numerical constant obtaine by evaluating the appropriate partial derivatives at the operating poir Under these conditions, the output of any block (represented by the arr pointing away from the block) is equal to the input to the block (repre sented by the arrow pointing towards the block) multiplied by the trans fer function. Each circle is a "summing point" whose output is simply the algebraic sum of the quantities which are represented by the arrows pointing towards it.

Notice that in this diagram all four feedback loops are clearly displayed. This makes it easy to decide which β's should be set equal to zero in order to interrupt a given set of loops while preserving in- tact the appropriate non-feedback pathways between a chosen input (cros hatched △) and a chosen output (solid △).

zero in order to open any given set of loops while preserving all of th direct pathways by which the chosen input can influence the chosen de- pendent variable.

III. How the Study of Magnification has Illuminated the Behavior of other Biological Systems.

Three further examples of using [Mag] to characterize the behavior of biological systems will be described very briefly. Details can be found elsewhere (Riggs, 1970).

(1) In adult mammals, the remarkable constancy of plasma calcium concentration has led to the notion that plasma calcium is under exquisitely precise negative feedback control. In fact, however, the change in plasma calcium concentration caused by a change in the rate of infusion of calcium is about one-half as great when the parathyroid hormone feedback loops are in operation as when they have been opened by thyroparathyroidectomy. This suggests that the major role of the parathyroid hormone is not to _regulate_ plasma calcium but to _elevate_ plasma calcium above the disastrously low levels which it assumes in the absence of hormonal control.

(2) It has long been known that pulmonary ventilation is controlled largely by the concentration of CO_2 in plasma. In studying this control system, most investigators have altered plasma CO_2 concentrations by adding CO_2 to the inspired air (an easily-manipulated input) rather than by changing CO_2 production in tissues (not so easily manipulated). Yet it turns out that the effectiveness of the chemical feedback regulation of ventilation as measured by [Mag] is very much greater when plasma CO_2 is increased by increasing CO_2 _production_ than when plasma CO_2 is increased by increasing CO_2 _inhalation_. It is not easy to convince classical pulmonary physiologists that this important difference exists!

(3) The body temperature of homoiotherms is indeed a variable exquisitely regulated by a variety of negative feedback loops. Yet lurking in the background is a positive feedback loop --- the "Q_{10} loop" --- which can threaten life itself when the regulatory loops have been suppressed, e.g., during anesthesia. The "Q_{10} loop" arises from the following cause-and-effect sequence: A decrease in ambient temperature causes a decrease in body temperature; a decrease in body temperature causes a decrease (measured by the "Q_{10}") in the overall rate of metabolic reactions; a decrease in metabolic rate causes a decrease in heat production; a decrease in heat production causes a still further decrease in body temperature. As it happens, this positive feedback loop is dangerously close to being unstable, though it is normally held

in check by the preponderant negative feedback loops. When those negative loops have been impaired by severe central nervous system depression, the malign influence of the "Q_{10} loop" may be displayed. Examples: a) the hyperthermia of "thyroid crisis" formerly seen when ill-prepared hyperthyroid subjects were anesthetized for subtotal thyroidectomy; b) the fortunately-rare syndrome of "malignant hyperthermia" during anesthesia (which seems to have a hereditary basis); and c) the well-known hypothermia of acute alcoholics who fall asleep in unprotected places during the winter months.

In Table 2 are listed estimates of $\lfloor Mag \rfloor$ for various biological systems under "normal" conditions. For comparison the corresponding values of $\lfloor OLG \rfloor$, the steady-state open-loop gain, are also given. The relationship between $\lfloor OLG \rfloor$ and $\lfloor Mag \rfloor$ is

$$[OLG] = \frac{1}{[Mag]} - 1 \qquad \text{for negative feedback loops}$$

$$[OLG] = 1 - \frac{1}{[Mag]} \qquad \text{for positive feedback loops}$$

(48)

Most of the values in Table 2 are rather crude approximations, subject to revision as better information becomes available. Indeed the major purpose of the table is not to catalogue current estimates of $\lfloor Mag \rfloor$ but to illustrate how much the effectiveness of feedback differs from one system to another.

IV. Concluding Remarks

Stear (1973) comments as follows on the need for more quantitative studies of negative feedback systems:

"Once the existence of negative feedback has been established it is important from a systems theory point of view to establish the (open) loop gain of the feedback system. The reason for this is that the loop gain provides a quantitative measure of the degree of regulation achieved by the feedback system; the higher the loop gain, the better the regulation. ...In view of the obvious physiological significance of loop gain as a measure of degree of regulation, ...it is surprising and somewhat disappointing to find so little attention given to the loop gains of those endocrine and metabolic systems which have been studied, despite the extensive characterization and discussion of them as feedback systems."

It is my hope that, by using techniques such as the ones described here, biologists will soon be able to gather the <u>quantitative</u> data which are needed to measure the performance of feedback systems under a variety of circumstances.

Table 2.
Approximate "normal" values for magnification, [Mag], and open loop gain, [OLG], for various biological systems

NEGATIVE FEEDBACK	[Mag]	[OLG]	Reference*
Body temperature (several loops)	1/400	399	R, p. 383
Plasma osmolality - antidiuretic hormone	1/100	99	R, p. 468
CO_2 - lung ventilation	1/30	29	R, p. 401
Allosteric control of synthetic paths	1/20	19	W, p. 195
Muscle blood flow - anoxia	1/4	3	R, p. 440
Plasma calcium - parathormone	1/2	1	R, p. 489
Glucose - insulin	5/8	3/5	R, Sec. 7.15
Pupillary reaction to light	5/6	1/5	R, p. 497
NO FEEDBACK			
Plasma Ca^{++} - calcitonin (adult human)	1.0	0	A, p. 1690
POSITIVE FEEDBACK			
O_2 consumption - work of breathing	1.01	1/100	R, p. 401
Iodide salvage from deiodination of thyroid hormone	1.5	1/3	R, Sec. 5.6
Body temp. - heat production	5	4/5	R, p. 383
UNSTABLE POSITIVE FEEDBACK			
(vicious circles)	Not defined (no steady state)	≥ 1.0	R, p. 500

*A = Aurbach and Phang, 1974; R = Riggs, 1970; W = Walter, 1975.

V. Appendix

(A.) Example illustrating various methods of calculating [Mag]

Given: A feedback system described by the two working equations

$$Y = f(X,U) = X^2 + U$$

$$X = g(Y,V) = -Y + V$$

where X and Y are loop variables in the steady state,
U and V are inputs, and U, V, X, and Y are all ≥ 0

1. Equations known. [Mag] calculated from its definition.

$$[\text{Mag}] = \frac{\partial X/\partial U \mid \text{closed}}{\partial X/\partial U \mid \text{open}} = \frac{\partial X/\partial V \mid \text{closed}}{\partial X/\partial V \mid \text{open}} =$$

$$\frac{\partial Y/\partial U \mid \text{closed}}{\partial Y/\partial U \mid \text{open}} = \frac{\partial Y/\partial V \mid \text{closed}}{\partial Y/\partial V \mid \text{open}}$$

Let us choose to work with the first of these.

Closed-loop situation.

Solve the working equations simultaneously for X.

$$X = -\frac{1}{2} + \frac{1}{2}\sqrt{1 - 4U + 4V} \qquad \text{, hence,}$$

$$\frac{\partial X}{\partial U}\bigg|_{\text{closed}} = -\frac{1}{\sqrt{1 - 4U + 4V}}$$

Open-loop situation.

We must open the loop in such a way that X will still depend upon
U. We must also have the same operating point as in the closed-loop
situation. With these restrictions in mind, the open-loop working equations are

$$Y = X_{op}^2 + U$$

$$X = -Y + V$$

where X_{op}, a constant, is equal to the original unperturbed operating-point value of X. (Note that Y is now uninfluenced by changes in X, so that the feedback loop is indeed open.)

Solving these equations simultaneously for X,

$$X = -X_{op}^2 - U + V \qquad\qquad , \text{ whence,}$$

$$\left.\frac{\partial X}{\partial U}\right|_{open} = -1$$

Thus, by the definition of [Mag],

$$[\text{Mag}] = \frac{-\dfrac{1}{\sqrt{1 - 4U + 4V}}}{-1} = \frac{1}{\sqrt{1 - 4U + 4V}}$$

2. Equations known. [Mag] calculated from the product of first derivatives.

By differentiating the closed-loop working equations,

$$\frac{\partial f}{\partial X} = 2X$$

$$\frac{\partial f}{\partial X} \cdot \frac{\partial g}{\partial Y} = -2X$$

$$\frac{\partial g}{\partial Y} = -1$$

Hence,

$$[\text{Mag}] = \frac{1}{1 - \dfrac{\partial f}{\partial X} \cdot \dfrac{\partial g}{\partial Y}} = \frac{1}{1 + 2X}$$

Substituting in the value of X found by solving the two closed-loop equations simultaneously,

$$[\text{Mag}] = \cfrac{1}{1 + 2\left[\cfrac{-1 + \sqrt{1 - 4U + 4V}}{2}\right]}$$

$$= \frac{1}{\sqrt{1 - 4U + 4V}} \qquad\qquad \text{, as before.}$$

3. Equations to be found experimentally.

(For simplicity the "observed values" of X and Y in this section will be errorless; i.e., they will actually be calculated from the closed-loop working equations.)

First Series of Experiments.

Manipulate Y via U, with V constant. Measure both X and Y. Plot X as a function of Y, and fit a curve to the points. The equation for the curve is an empirical estimate of X = g(Y,V).

	INPUT	OBSERVED STEADY-STATE VALUES	
Let V = 8	U	Y	X
	8	8	0
	6	7	1
	4	6.4384	1.5616
	2	6	2
	0	5.6277	2.3723

See Figure 5 for a graph of these points.

Second Series of Experiments.

Manipulate X via V, with U constant. Measure both X and Y. Plot Y as a function of X and fit a curve to the points. The equation for the curve is an empirical estimate of Y = f(X,U).

	INPUT	OBSERVED STEADY-STATE VALUES	
Let U = 4	$\underline{V}$	$\underline{X}$	$\underline{Y}$
	4	0	4
	6	1	5
	8	1.5616	6.4384
	12	2.3723	9.6277
	20	3.5311	16.4689

See Figure 5 for a graph of these points.

Figure 5

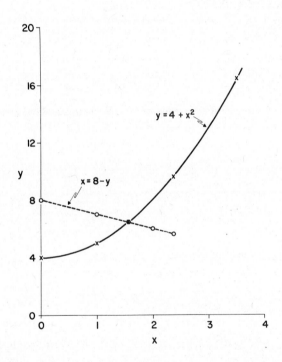

Plot of $X = 8 - Y$ and $Y = X^2 + 4$. The operating point is at the intersection of these curves where $X = 1.5616$ and $Y = 6.4384$ ($U = 4$ and $V = 8$).

The two curves thus found:

$$X = 8 - Y \; ; \; Y = 4 + X^2 \quad ,$$

intersect at the operating-point values of X & Y obtained by solving the two equations simultaneously:

$$X = 1.5616 \; ; \; Y = 6.4384$$

(These are the values appropriate for the chosen constant inputs, U = and V = 8.)

Using the product-of-first-derivatives method:

$$[\text{Mag}] = \frac{1}{1 - \frac{\partial f}{\partial X} \cdot \frac{\partial g}{\partial Y}} = \frac{1}{1 + 2X} = \frac{1}{1 + 2[1.5616]}$$

$$= 0.2425.$$

Hence at this particular operating point negative feedback has reduced the effect of an external disturbance to about 24% of what it would ha been without feedback.

To find [Mag] at any other operating point, a new series of experiments would have to be performed.

(B) Using Taylor's series to linearize around a particular point can best be illustrated by a simple example in which a dependent variable, $y(x)$ is a function of only one independent variable, x. Let $[x_o, y_o]$ be the point about which we are to linearize the function $y(x) = f(x)$. The Taylor's series expansion of $f(x)$ is

$$f(x_o + \Delta x) = f(x_o) + \frac{\Delta x}{1!} f'(x_o) + \frac{[\Delta x]^2}{2!} f''(x_o) +$$

$$\frac{[\Delta x]^3}{3!} f'''(x_o) + \ldots$$

where Δx is an increase in x, usually taken to be a small increase, and all of the derivatives are to be evaluated at the point $[x_o, y_o = f(x_o)]$. Neglecting all second- or higher-order terms in Δx, and rearranging slightly, we obtain the linear approximation

$$f(x_o + \Delta x) - f(x_o) \doteq f'(x_o)\Delta x$$

But the left side of the above equation is simply Δy, the change in y caused by Δx, the change in x.   Hence

$$\Delta y \doteq f'(x_o)\Delta x$$

Note that for vanishingly small changes, this approximation becomes

$$\frac{dy}{dx}\bigg|_{x \,=\, x_o} = f'(x_o)$$

which is true by definition.

As a numerical example, suppose that

$$y(x) = f(x) = \frac{2}{3x^2}$$

Suppose we wish to linearize around the point where $x_o = 2$ and $y_o = 1/6$. The first derivative of y with respect to x is $-4/3x^3$.   Evaluating this at $x_o = 2$, we have $f'(x_o) = -1/6$.   Hence

$$\Delta y \doteq [-1/6]\Delta x$$

[Check: Let $\Delta x = 0.01$. Then

$$y(2.01) = 2/3(2.01)^2 = 0.16501242,$$

whence $\Delta y = 0.16501242 - 0.16666667 = -0.00165425.$

Our linear Taylor's series approximation gives

$$\Delta y \doteq [-1/6] \, [.01] = -0.00166667$$

which is well within 1% of the correct value.]

(C.)   For several reasons, this is a weak and unrealistic assumption, made chiefly for didactic simplicity.   In reality, food would be consumed faster initially when the load is heavy.   On the other hand, progress along the trail would be slower initially, so that more time would be spent with a heavy pack.   If due account were taken of these matters, W_f would have to be treated as a time-dependent variable.   But we cannot deal with variables which change with time when we are confining attention to the steady state. In steady-state analysis, therefore, it is sometimes necessary to replace a quantity which, in fact, varies with time by its mean value which does not.

A more serious defect of equation (19) is that the maximum load allowed by (17) for a nonzero velocity will be the sum of the weight of equipment and <u>one-half</u> of the true maximum weight of food. This defect will probably be important only for very large loads.

It might be more realistic to replace (19) by some such equation as

$$\overline{W}_f = h(W_{f.max})W_{f.max}$$

where

$$h(W_{f.max}) = 0.5 \left[1 + \exp \left[-0.2(\frac{v_{max}}{k} - W_{equip} - W_{f.max}) \right] \right]$$

The "fudge factor", $h(W_{f.max})$, would be practically equal to 0.5 for all reasonable loads, but would rapidly rise towards unity as $W_{f.max} + W_{equip}$ approaches v_{max}/k, the maximum load allowed by (17).

Ⓓ Let

$$z_i = g(z_1, z_2, \ldots, z_{n-1}; I_1, I_2, \ldots, I_m)$$

be the equation for z_i, the <u>i</u>th dependent variable, as a function, g, of the n-1 other dependent variables, z, and of the m independent variables, I. Then

$$\Delta z_i = \frac{\partial g}{\partial z_1} \Delta z_1 + \frac{\partial g}{\partial z_2} \Delta z_2 + \ldots + \frac{\partial g}{\partial z_{n-1}} \Delta z_{n-1} + \frac{\partial g}{\partial I_1} \Delta I_1 +$$

$$\frac{\partial g}{\partial I_2} \Delta I_2 + \ldots + \frac{\partial g}{\partial I_m} \Delta I_m$$

is the corresponding linear equation for small deviations. Each Δ in these equations represents a small deviation of the specified variable from its operating-point value, and all partial derivatives must be evaluated at the operating point previously found. Similar equations must be written for each of the i independent variables.

VI. References

Aurbach, G. D. and Phang, J. M. (1974). "Vitamin D, parathyroid hormone, and calcitonin". Chapter 70 in Medical Physiology, Vol. 2, 13th Edition (Mountcastle, V. B., Editor) C. V. Mosby, St. Louis.

Riggs, D. S. (1970). "Control Theory and Physiological Feedback Mechanisms". Williams and Wilkins, Baltimore.

Stear, E. B. (1973). "Systems theory aspects of physiological systems". pp. 496-500 in Regulation and Control in Physiological Systems, (Iberall, A. S. and Guyton, A. C., Editors), Proceedings of a Conference. International Federation of Automatic Control. (Distributed by the Instrument Soc. of America.)

Tucker, V. A. (1975). "The Energetic Cost of Moving About". American Scientist 63, 413-419.

Walter, C. (1975) in "Enzyme Reactions and Enzyme Systems". Enzymology Series, Number 4. Marcel Dekker, New York.

From complex, single organisms to ecosystems -- Dr. Goldstein's contribution to the volume addresses the addition of abiotic factors to mathematical models, categorizing them by level of resolution and by objective. Based on his own extensive experience, he describes difficulties associated with ecosystem modeling, challenging mathematicians and biologists to confront them in an effort to increase the value of mathematical models as tools of biological discovery.

REALITY AND MODELS: DIFFICULTIES ASSOCIATED WITH
APPLYING GENERAL ECOLOGICAL MODELS TO SPECIFIC SITUATIONS[1,2]

Robert A. Goldstein
Electric Power Research Institute
P.O. Box 10412, Palo Alto, CA 94304 USA

This paper presents a personalized and philosophical evaluation of some present capabilities and limitations of ecological modeling. I stress that the evaluation is personal and based primarily upon my own research and experiences. The paper is not meant to be a state-of-the-art review, hence there are few references to the literature.

Modeling A Multifaceted Activity

Oftentimes, comments or criticisms are made regarding ecological modeling that imply ecological modelers are a homogeneous group in terms of philosophical approaches to research and modeling methodologies employed. In truth, a considerable diversity of philosophies exists among modelers and there is a large variety of modeling techniques and approaches that can be used in ecological research. It should be recognized that there are as many different types of modelers as there are flavors of ice cream in the franchise ice cream parlor and that any judgment made as to the potential value of ecological modeling should not be based on the ideas of one modeler.

There are differing opinions regarding the role modeling and modelers should play in ecological research. How should modeler and experimentalist interact? (Note, modeling and experimentation are not mutually exclusive activities, and a single individual can be engaged in both.) Does the modeler develop a model, for which the experimentalist collects data to validate? Does the experimentalist collect data, for which the modeler constructs a model to "fit"? Or do the modeler and experimentalist design a research project that welds modeling and data collection? In either case, who is chief? Does there need to be a chief?

[1] This paper is based on research conducted by the author while an employee of the Environmental Sciences Division, Oak Ridge National Laboratory, operated by Union Carbide Corporation for ERDA.

[2] The contents of this document are the responsibility of the author, and should not be attributed to EPRI.

There are several classes of ecological models and some modelers have a distinct preference for using a particular class. Ecological models are often referred to as "ecosystem," "process," or "population" models. These terms are not well defined; hence, several individuals may disagree as to what term best describes a single model. In fact, most models probably combine elements of different classes. Population models focus on the dynamics of one or more populations. These models incorporate basic population processes, such as birth, natural mortality, predation and competition. The dependent variables are population densities, which can be measured either in terms of density of individuals or biomass. Historically, these models rarely have incorporated, explicitly, environmental factors such as temperature, humidity, and solar radiation. This was a result of the models having to be kept relatively simple, since the models were studied by analytic and graphical techniques. As the use of computers has become widespread, population models have been becoming more complex and the inclusion of environmental factors has become more common. Although population modeling is applicable to both animal and plant populations, it has been applied primarily to animal populations.

Process models focus on the dynamics of major physio-ecological processes such as photosynthesis or thermoregulation. These models tend to incorporate, explicitly, basic physiological, physical and chemical mechanisms that govern the processes, and environmental factors (e.g., temperature, humidity, solar radiation) which influence these mechanisms. The dependent variables tend to be energy, biomass, water, and nutrients being accumulated and transferred. Process models are most frequently applied to the study of vegetation.

Ecosystem models focus on the integrated dynamics of an entire community or ecosystem. They describe the flow of energy, biomass, water and nutrients through the system. Because of the breadth of their scope, ecosystem models do not contain the detailed physiology, chemistry and physics that process models do. Ecosystem models also tend to have coarser temporal and spatial resolutions than process models. Ecosystem models usually contain elements of both population and process models.

A wide variety of mathematical constructs have been used in ecological modeling. A diversity of mathematical approaches is more likely to provide greater insight into the behavior of a complicated system than would any single approach. One would certainly not attempt to disassemble a car's engine with a single screwdriver. In certain instances, controversies have arisen concerning the contrasting validity or inherent value of pairs of approaches; e.g., linear versus nonlinear,

stochastic versus deterministic, difference versus differential. In
any given situation, the choice of mathematical construct should be
strongly governed by the specific research objectives. Increased mathematical sophistication should not be presumed to necessarily provide
better results.

Summarizing my personal philosophy regarding the role of mathematical modeling in ecology, I feel that modeling should be an important
component of any ecological research program, but that modeling by itself does not constitute a research program. Models are research tools.
The types of models and mathematics employed in a research program should
be based on the specific objectives of the research.

Brief Historical Background

Historically, mathematical ecological modeling and theory was
oriented towards population dynamics. Most mathematical ecological
modeling was based on the equations of Lotka (1925) and Volterra (1928)
or modifications of these equations. Discrete population models drew
heavily on the work of Leslie (1945). During the nineteen sixties
there was an evolution in the ecological sciences that resulted in increasing research emphasis on the development and application of process
and ecosystem mathematical models.

The current vitality in process modeling appears to trace its gene-
sis to the research of Monsi and Saeki (1953) on primary production in
forests. There appears to be a gap of almost a decade before Europeans
and Americans started building on the Japanese work. Up to the late
sixties, agricultrual scientists, such as Duncan (1967) and deWit (1965),
clearly led ecologists in developing and applying process models.

During the sixties, the development and application of ecosystem
models derived considerable impetus from the publications and lectures
of Olson, Holling, Watt, Goodall, Van Dyne and Patten. The International Biological Program played a major role in intensifying process
and ecosystem modeling research.

The primary objective of most of the earlier research of which process and ecosystem modeling was a major component was to increase basic
understanding of natural ecosystem structure and dynamics. Of course,
population models have an extensive history of application to practical
problems in the disciplines of entomology and marine fisheries. Recently, there has been considerable interest in ecological mathematical
modeling in connection with studying problems associated with environmental assessment and increasing agricultural production.

Objectives of Ecological Modeling

Within a given ecological research program, modeling can be a valuable procedure in helping to address a number of frequently occurring research objectives. A basic research objective is increased fundament understanding of the system being studied. This need not be an objecti of all research programs. Oftentimes, there is a desire to produce a specific output from a given system. In many circumstances, this goal can be achieved through a well-designed manipulation of the system's inputs, without any attempt to derive a basic understanding as to how the system functions. This type of approach is frequently referred to as an "input-output" analysis and the system is described as a "black box." Mathematical modeling techniques can be very helpful in this typ of analysis as well.

Other frequently occurring research objectives for which the appli cation of modeling is valuable are synthesis of extant knowledge and un derstanding of a system, critical analysis of hypotheses about a system functioning, identification of fundamental constraints on a system's functioning, and identification of mechanisms to which a system's behav ior is most sensitive. The formulation of a model at the initiation of a research program provides a valuable framework for organizing existin knowledge and integrating the efforts of individual investigators worki on the program. The formulation of a model also leads to a review of current hypotheses concerning the system's functioning.

Since a model is based on specific hypotheses about the system's behavior, manipulation of the model using data in the literature and data forthcoming from the research program will enable the validity of the hypotheses to be tested. There are two types of validity tests. The first is that the individual hypotheses that form the model are con sistent with one another; that is, the model is composed of a self-con-sistent set of hypotheses. Mathematical analysis of the model is a valuable approach to checking for logical or conceptual inconsistencies The second type of validity test is comparing the model's behavior with available data for consistency.

Analyses to determine factors to which a system's behavior is most sensitive and analyses to identify fundamental constraints on a system' behavior can be applied to both properties inherent to the system (e.g. stomatal resistance of plants or natural mortality of fish larvae and eggs) and properties of the environment (e.g., precipitation patterns or elemental deposition through rainfall and dryfall). While sensitivi analyses are often performed in connection with model development and application, constraint analyses are very rare.

Within a research program, modeling aids in the identification of additional research needs. Within a given system, modeling can help identify processes where increased basic understanding is needed. For example, in formulating a model for studying effects of acidic deposition on forest growth and development, it would become apparent that current understanding of elemental movement from soils into roots is inadequate. Application of models lead to identification of information gaps, in the available data, that need to be filled in order to evaluate important parameters.

As mentioned in the previous section, over the next decade, modeling should become a very important means of addressing questions related to environmental assessment of perturbations and management strategies applied to ecosystems. Mechanistic based process models have greater potential in this area than either ecosystem models or simple nonenvironmentally dependent population models. Abiotic factors (e.g., temperature, moisture, solar radiation) that are important in controlling ecological processes need to be explicitly included in the models that are applied to environmental assessment questions, since in most situations it will be an alteration of the abiotic environment that will have the most pronounced and long lasting impact on the system.

Difficulties Associated with Ecological Modeling

A mathematical model of a system attempts to depict, relevant to specific objectives, essential features of the system. If this attempt is successful, the model can be considered realistic. However, the fact that a given model is realistic, does not necessarily imply that the model may be applied to an arbitrarily chosen real system in such a manner that the behavior of the model matches, according to some set of independent criteria, the behavior of the real system. Over the last decade, ecological scientists have made considerable progress in developing realistic mathematical models of ecological phenomena and systems. However, several major factors exist which limit the ability of these models to predict what will be the effects of a perturbation of a specific real system. The difficulties associated with the application of mathematical modeling to ecological studies are not necessarily unique to ecology, nor do they negate the potential value of ecological models to meet the objectives discussed in the previous section; however, these difficulties should not be glossed over by modelers. The difficulties identify research topics that mathematical and theoretical ecologists should be giving considerable attention to, and include lack of fundamental unifying theories, sparse data sets, biological varia-

bility, spatial heterogeneity, and interactions of processes operatin
at significantly different temporal and spatial scales.

Ecology, as a science, is in its early stages of development rela
tive to disciplines such as physics and chemistry. As such, ecology
has not as yet formulated fundamental unifying theories that can be
used to deduce quantitative ecological behavior. Ecology has a number
of loosely defined principles concerning concepts such as succession,
competition, diversity and stability. However, these principles tend
to be qualitative and there exists considerable uncertainty concerning
their validity.

As a result of ecology's lack of fundamental unifying theories,
ecological models most often take the form of simulations which are
complex composites of known and hypothesized relationships, from which
dynamic properties of the system being studied are induced. Simulatio
modeling has several inherent weaknesses. Simulation models tend to
possess very large numbers (usually scores) of parameters, and hence
very high degrees of freedom. Thus it is frequently easy to simulate
observed patterns of behavior in the real system, regardless of the
model's conceptual validity. This is especially true for the case of
complex systems, which have not been intensively monitored over a broa
range of environmental conditions. In fact, for many situations, ther
are probably multiple sets of parameter values that would produce simulations that are in equivalent agreement with the available data. A
strength of mechanistic process models is that each parameter has some
physical, chemical or physiological interpretation, and can frequently
be estimated independently of most of the other parameters. Even if
independent estimation of the parameter is not possible, usually bound
can be placed on the possible values the parameter could have. An example of a mechanistic process model is an evapotranspiration model,
where most of the parameters would represent water conductivities, wat
potentials, and heat and vapor transfer properties of both soil, litte
and vegetation.

Another inherent weakness in simulation modeling involves the application and development of the models. For a given research project
the initial model is formulated according to a conceptualization of ho
the system functions. The initial model is compared to data taken fro
the real system, and modified to improve agreement between simulation
and observed behavior. Over a period of time, as study effort intensi
fies, model modification and data collection proceed in an iterative
manner until there is satisfactory agreement between simulation and sy
tem behavior for all situations of interest. Since the construction o
the model is almost totally an inferential process, there is a potenti

anger that after the model has undergone many modifications, it no
onger presents a strong (i.e., well integrated composite of relation-
nips) and clear conceptual framework of the system's structure and dy-
amics. Thus one can start with a solid conceptual framework and rapidly
nd unwittingly dissolve it as an·attempt is made to modify the model
o simulate new data. Of course, the potential for this is greatest
hen one group attempts to apply a simulation model developed by another
roup. A mechanistically derived simulation model would to some extent
itigate the potential danger. The existence of a broad theory that en-
ompassed all of the relationships in the model would limit the avenues
or modification and lessen the potential of dissolving the conceptual
ramework.

A second difficulty associated with ecological modeling is what I
efer to as sparse data sets. It is difficult to validate a generic
odel for a given ecological phenomenon (e.g., phytoplankton primary
roduction) or system (e.g., forest growth and development) because of
 lack of complete data sets in the ecological literature. There will
xist many data sets relevant to a given phenomenon or system, but few,
f any, will contain measurements of all the extrinsic and intrinsic
actors which are important in controlling the dynamics. It is also
ifficult to apply an ecological model to a specific situation (be it
 location, community, or species), because most likely the data will
ot exist to quantify all of the model's parameters for that specific
ituation. Oftentimes, histories are available for particular outputs
f a system. For instance, stream flow data, which can be compared to
he output of a watershed hydrological model, are available for many
ites, but the data needed to check water contents at different soil
epths, which are also simulated, are available for only a few sites.
t should be noted that long periods of time are usually required to
ollect the data needed to estimate the parameters for or validate an
ecological model.

Biological variability is another difficulty. All protons placed
in identical environments will respond identically. However, all _Sym-
phurus plagiusa_ placed in identical environments will not respond identically. Differences in biological responses between individuals of the
same species, or representative samples of the same community can be
based on genetic differences, environmental preconditioning differences,
or combinations of the two. Biological variability poses special problems to ecologists attempting to develop and apply deterministic models.

Spatial heterogeneity in the environment conditions biota to respond differently to identical situations, and hence produces a spatially dependent biotic response surface. Therefore, considerations of

spatial heterogeneity and biological variability are connected in some
aspects. Spatial heterogeneity can be an important factor influencing
system dynamics and stability. Spatial heterogeneity presents a conceptual difficulty to ecologists attempting to develop and apply point
models, i.e., models where spatial dependencies have been removed by int
grating the spatially dependent factors over all of space. The proble
is how to best integrate or average out the spatial dependencies. Explicit treatment of spatial heterogeneity presents difficulties also;
since it drastically increases the complexity of the model.

In conducting ecological research, one is confronted with a large
array of interacting phenomena, whose fundamental dynamics conceptuall
lie within an extremely broad range of temporal and spatial scales of
resolution. The phenomena range from the biochemical, to the physiolo
ical, to the ecological, to the landscape, and finally to the global.
There is a corresponding spatial scale which goes from the cell, to th
individual organism, to the community, to the watershed, to the geographical region. The time scale spans from seconds to centuries. It
is of questionable value to attempt to construct a single model that
encompasses the entire hierarchy of processes, or to devote all of one
energies to a single level of the hierarchy. The best procedure is to
view the system from several levels of the hierarchy simultaneously, b
developing and applying a suite of models, each appropriate to a given
spatial-temporal level of resolution.

The final difficulty that I wish to discuss involves the inherent
nonlinearity of ecological phenomena and systems. It is common to fin
nonlinear interactions dominating the dynamics of ecological phenomena
and systems. The classical example is the predator-prey interaction.
Both threshold and saturation reactions occur commonly in ecological
systems. Nonlinear models, in contrast to linear models, pose special
problems as regards mathematical analysis. The capability to derive
analytical, in addition to numerical, solutions to a mathematical model
is highly valuable. Analytical procedures give greater insight into
the functioning of the system, than do numerical procedures. Analytica
procedures are valuable tools for checking the model for logical or
conceptual inconsistencies. Although nonlinear interactions make the
mathematical analysis of a model more complex, they frequently increase
the stability of a system and hence simplify the sytem's dynamical response to perturbation.

Conclusion

The mathematical modeling of ecological phenomena and systems is still in an early stage of development. Despite the difficulties associated with ecological modeling, there has been considerable progress in integrating mathematical modeling into ecological research and in developing general ecological models. Over the next decade, I expect this trend to continue and I further expect to see the widespread application of models to problems of environmental assessment.

References

Duncan, W. G., R. S. Loomis, W. A. Williams, and R. Hannan. 1967. A model for simulating photosynthesis in plant communities. Hilgardia 38, 181-205.

Leslie, P. H. 1945. The use of matrices in certain population mathematics. Biometrika 33, 183-212.

Lotka, A. J. 1925. Elements of Physical Biology, Williams & Wilkins, Baltimore.

Monsi, M. and T. Saeki. 1953. Über den lichtfaktor in den pflanzengesellschaften und seine bedeutung fur die stoffproduktion. Jap. J. Bot. 14, 22-52.

Volterra, V. 1928. Variations and fluctuations of the number of individuals in animal species living together. J. Cons. Int. Explor. Mer. III, Vol. I. Reprinted in: R. N. Chapman (ed.), Animal Ecology, McGraw-Hill, New York. 1931.

DeWit, C. T. 1965. Photosynthesis of leaf canopies. Agr. Res. Rep. 663, Inst. Biol. Chem. Res. on Field Crops and Herbage, Wageningen, 57 pp.

In the next paper Professor van der Vaart examines the components of a successful interaction between mathematics and biology. In this paper he suggests criteria for identifying contributions to biological discovery and provides examples in which mathematical models have contributed to our understanding of biology or led us astray.

BIOMATHEMATICAL MODELS: SOME TRIUMPHS AND
SOME DEFEATS

H. R. van der Vaart
Department of Statistics
North Carolina State University

Several authors in this volume give detailed discussions of contributions made by mathematical methods to biological discovery: e.g.,
theory of systems that are far from equilibrium; enzyme systems; feedback systems; computer simulation providing a glimpse of the whole after
the parts have been analyzed; computer simulation helping to determine
environmental policy; the vigorous development of neurophysiology after
the Hodgkin-Huxley equations appeared on the scene. In view of all this
one would think that the conclusion that mathematical methods can, indeed, contribute to biological discovery is safely entrenched, so that
there is no need for any further discussion.

On the other hand, it is fair to say that there are aspects of the
interaction between mathematics and biology that are not told by the
success stories: these aspects are uncovered only when one looks into
the beginning stages of the development of a new result in biology, not
when one gazes at the finished product. I will cite some examples of
such beginnings, some anecdotes if you will, that in my personal opinion
give significant insights into what can go right or what can go wrong
between mathematics and biology.

It is hard to categorize all possible occurrences in this area,
but it is helpful to have at least a partial grasp on their multiformity.
It is, then, good to keep the following scheme in mind when evaluating
the failure or success of any attempted cooperation between mathematics
and biology:

1) the biological competence of the mathematician (which includes
his capacity for listening and talking to his biological teammate, if
any, and his ability to absorb what things biological he hears or reads);

    a) for any work in 'fundamental' biology this will influence
the biological content of the resulting theory;

    b) for any work in 'applied' biology this will influence the
practical usefulness of the resulting mathematical formalism
(such as predictor equations, optimization methods, etc.);

2) the mathematical competence of the biologist (which includes
his ability as an expositor of biological problems to his mathematical

teammate, if any, and his aptitude to keep a tight rein on the mathematical phantasies of his teammate, as well as his capacity for listening
to, reading, and interpreting mathematical results);

   a) for any work in 'fundamental' biology this will influence
      the appropriateness, or lack of it, of the mathematical
      structure as a reflection of biological insight;
   b) for any work in 'applied' biology this will help, for in-
      stance, select the most sensible statistical recipe (e.g.
      avoid linear regression formulae when the regression is
      clearly curvilinear).

Connected with these two points are such aspects as:

   3) the selection of the object of research;
   4) the strategy of attack on a problem when selected;
   5) the main thrust of the research, the level of resolution, the
amount of detail developed in the proposed answer to the proposed prob-

   In this scheme we have not explicitly mentioned such obvious nec-
essities as the biological competence of the biologist, or the mathematical competence of the mathematician. Our present concern is with some
typical aspects of the interaction between mathematics and biology, not
with the inner workings of either field on its own.

   Since our concern here is primarily with the performance of bio-
mathematical models as contributors to biological discovery, it would
be helpful also to analyze the possible meanings of this term, 'biological discovery'.

   Now biology is unmistakably oriented towards experiments and obse-
vations rather than towards theory. Some biologists would go so far as
claiming that where there are no experiments or observations, there is
no biology. This line of reasoning leads to the denial of the status
'contributor to biological discovery' for any activity that does not
originate new biological experiments or observations. By the same token
any activity that does originate such experiments, at least when these
experiments are accepted as 'significant' by a significant number of
biologists, should obtain the status of 'contributor to biological discovery'. So should any activity that leads to new ways of handling experiments, whether these new ways consist in new instruments to conduct
them with, or in new methods to design or analyze them with, provided
these instruments or methods are deemed improvements over what was avai
able before. According to these criteria Darwin's Theory of Evolution
(which, of course, is nonmathematical in itself) has clearly been a
major contributor to biological discovery: decades of work in compar-
ative anatomy and population genetics, to name just two fields, were

inspired by Darwin's theory. But also the arsenal of statistical methods of design and inference should be counted as a major contributor to biological discovery: not only did R. A. Fisher develop a good number of them in constant interaction with agricultural pursuits, but also is the number of such research projects in the life sciences as include some application of statistics almost countless. Statistical methods, of course, are of mathematical parentage.

In a broader sense, almost any experiment is preceded by some kind of thinking; so thinking is a contributor to biological discovery -- as indeed one would expect or hope. Whenever this thinking process has some mathematics in it (as it might, whenever the thinking is done by a mathematically competent person), the mathematics makes a contribution.

There is a strange aspect to biological discovery (as there is to physical discovery, chemical discovery, and many others). In the above I have used the words 'new' and 'significant'. An experiment is 'new' when it has not been done before, and it is deemed 'significant' if enough people of consequence acclaim it. The former concept, '<u>new</u>', in spite of the apparent simplicity of its definition, is utterly vague: on the one hand, when an experiment that was previously conducted at 35° C, is now being conducted at 35.5°C, this will in general not be regarded as new (unless the outcome is unexpectedly different); on the other hand, among the experiments that are recognized as new, there are plenty that do have several features in common with some earlier ones. The latter concept, '<u>significant</u>', is even more treacherous. As an example, consider this story: in 1874, before obtaining his doctorate, J. H. van't Hoff published an 11-page paper (in Dutch) containing the theory of the asymmetric carbon atom, thus opening the way to an understanding of stereoisomers and optically active compounds. Next year, this paper appeared in French (J. H. van't Hoff, 1875), and in 1876 it was translated into German because one man of consequence, J. Wislicenus (professor in chemistry at Würzburg) thought it to be of "epoch-making significance". However, another man of consequence, H. Kolbe (professor in chemistry at Leipzig), justly famous for his outstanding synthetic work (the first to synthesize the organic compound acetic acid from its elements) greeted this publication with an unusually vitriolic criticism (see pp. 6-8 in vol. 1 in the Springer series "Molecular Biology, Biochemistry and Biophysics" (1967), which consists mainly of a translation of J. H. van't Hoff's inaugural address: Imagination in Science). Kolbe being the older and more famous man, one sees that even contributions that by historical hindsight have been proven to be of extreme significance, may originally have a hard time being recognized as such. For

another example, contemplate the following: in 1859 the president, Th
Bell, of the Linnean Society of London, when reviewing the papers read
before the Society during the preceding year, regretfully noted that
there had not been 'any of those striking discoveries which at once
revolutionize, so to speak, the department of science on which they
bear' or which 'shall produce a marked and permanent impress on the
character of any branch of knowledge'. Yet, on 1 July 1858 Ch. Darwin
and Alfred Wallace had read to the Society their "On the tendency of
species to form varieties" (see preface of Bell, 1959). Also, the wor
of Mendel (1865) in genetics, as we know, remained virtually unnoticed
for about half a century: there was nobody of consequence who as an
expert in genetics could pronounce that work significant because Mende
himself was the only geneticist in existence. On the other hand, most
authors of dissertations without a 'cum laude' predicate do go without
a Nobel prize, and many papers that are buried in old journals may as
well stay where they are.

So when judgment is to be passed on the question whether mathematical methods have contributed to biological discovery, we have to worry
among other things, about the possibility that a major contribution has
in fact been made, but is not recognized as such by the presently avail
able biological talent, as well as about the possibility that a certai
contribution has been pronounced significant, but will meet with the
indifference of a later generation. In the first case a mathematical
method will fail to receive its due recognition; in the second case its
undue recognition will have to be retracted in the future.

Now let us look at some examples (for more details see a forthcoming publication by the present author in the SIAM Review).

<u>Growth</u> <u>curves</u>. One of the early apparent successes of mathematics
in biology consisted in the generation of growth curves: mathematical
expressions that were used to represent the size of a growing organism
a function of time (e.g., exponential, logistic or autocatalytic, Gom-
pertz, L. von Bertalanffy), or the corresponding sizes of two developin
organs (allometric formulae). Often a reasonable fit was obtained. So
the question arises: why then was A. E. Needham in his 1964 book (see
his preface, p. v) quite unhappy with these early successes of the math
ematical description of growth? It is a matter of the above points 4
and 5, and to some extent 1: the strategy of attack was not a thorough
analysis of the entire growth process, but rather a brief, superficial
argument; the main thrust of the research was concerned with just one
aspect of growth: the overall size of the organism (or for the allomet
formulae, the overall size of two parts of the body); consequently, the

biological content of most of these growth formulae was meager. With regard to this last statement it is remarkable that most of the work with those growth formulae was done by card-carrying biologists rather than mathematicians: the successful fit of the overall growth phenomena by the formulae seems to have fooled them into not realizing that they did not really understand the phenomenon of growth, e.g., as a cellular process (cf. the above point 5 on the level of resolution). In Needham's words: "Because growth studies began with relatively quantitative methods, this tended to obscure the need for an orthodox analysis", namely at the level of cytology, endocrinology, nutrition, etc. Nowadays the only use of these formulae is for purely descriptive purposes, e.g., in fishery research when one wants to do something about correcting for the difference in weights between fish of different ages.

Another example where the mathematical model has not done the biology much justice, is provided by a "mathematical model of the chemistry of the external respiratory system", which was constructed by some people of the RAND Corporation and published in a number of technical reports and articles, with a more definitive publication in the Fourth Berkeley Symposium. Without going into much detail the essence of this model is as follows. The model studies three compartments: 1 the air in the lung-sacs, 2 the bloodplasma, and 3 the red cells in the blood. The model lists chemical compounds of interest in the respiratory process (partially different compounds in the different compartments), and it seeks to determine theoretically the concentrations of these compounds in the three compartments. Among the resulting values some are good, others poor. The ratio of the concentrations of CO_2 and H_2CO_3 is off by a factor of 10^{14}. What is the reason for this lack of success? It would seem to come mostly from the above points 4 and 1. The approach to the problem is based on a result from the equilibrium thermodynamics of closed systems. However, the lung-blood system is not closed and is not in equilibrium. Indeed, the very essence of the respiratory process defines the system as open: pick up oxygen in the lungs and loose it in the body; pick up carbon dioxide in the body, and loose it in the lungs. Now the result from equilibrium thermodynamics (minimization of Gibbs free energy), which forms the foundation of their treatment, is invalid in non-equilibrium thermodynamics. There are other, more mathematical, difficulties, which will be discussed elsewhere.

Now as opposed to these less successful interventions of mathematical methods in biological problems there are a number of examples of outstanding success. There is the case of genetics. Right from the start, with Mendel himself, genetical thinking was laced with mathematics, and this trend has continued unto this day, both in population genetics (e.g., see Kimura & Ohta, 1971, and the journal Theoretical Population Biology) and in the study of heredity in the individual organism, especially with respect to such phenomena as linkage and crossing-over, and the ensuing chromosome mappings (e.g., Bailey, 1961). Not surprisingly, statistical methods loom big in genetical research (e.g., Eland Johnson, 1971; Kempthorne, 1957). All of this has obviously contribute to genetical discovery.

Demography is unthinkable without mathematics. From the old days when Lotka was canonizing the concept of stable age distribution and setting up his renewal-type integral equation for the overall birth rat (e.g., see Lotka, 1939, where older literature is quoted), during the later years when Leslie, preceded by others, published the discrete model for the study of age distributions (for an example of demographic use see Keyfitz, 1967), to the present time, when the models allow vita rates to vary (Lopez, 1961; Coale, 1972), other features than numbers in age classes are studied (Sheps & Menken, 1973), and increasing mathematical sophistication is introduced (e.g., see Hoem, 1972), mathemati has played a vital role in the handiwork of demography. Offshoots into the study of animal populations (which were, in fact, Leslie's primary concern) should be mentioned, for instance the study of optimal harvesting (of which an elementary example is in Usher, 1972). Finally, i is to be hoped and expected that mathematical models will play a vital role in predicting consequences of contracting birth rates (a taste of this in Keyfitz, 1972), thus helping to avoid such disasters as followe the unfortunate drive, by the mathematical community, for more and bigg Ph.D. programs (in mathematics) in the late sixties, just a few years before the bottom fell out of the mathematical job market.

The kinetics of enzyme action is another field which is inextricab. intertwined with mathematics (e.g., see Laidler & Bunting, 1973). Just one aspect of the relevant theory, viz. the Michaelis constant, has bee the center of attention in much experimental work; and the same is true of many other parts of the theory.

My personal favorite example of a mathematical model contributing to biological discovery consists of the Volterra equations for two competing species. They showed that under certain conditions regarding the parameters in those equations two competing species could not co-exist for an indefinite period of time. Interestingly the mathematician Volterra did not come up with these equations because he was looking for an application of his mathematical talents, but because a live biologist (his son-in-law U. d'Ancona) came to him with a fisheries problem.

Although the equations were quite crude and relatively little biological insight was incorporated in them, they did lead to an enormous amount of experimental work. First in line was Gause (1934), whose work was explicitly provoked by Volterra's publications; he added experimental evidence to Volterra's mathematics, and the above idea (of the non-coexistence of two competing species) became known as the 'competitive exclusion principle' (sometimes called Gause's principle). The amazing thing about this concept is that it was originated by a mathematical model crude beyond recognition, but capturing what turned out to be a central question in the concept of inter-species competition. The ensuing biological experiments led in turn to an analysis of the concept of competition, to the discovery that with relatively small changes in environmental conditions (such as temperature) the formerly dominant species may become the loser, and to a host of other discoveries: the importance of genetical differences, of partitioning the physical space into different habitats, of inhomogeneous environments, and many other things. And all of this because of two little old equations; not bad as a contribution to biological discovery!

BIBLIOGRAPHY

Bailey, N. T. J. (1961), Introduction to the mathematical theory of genetic linkage; Oxford, Clarendon Press, x + 298 pp.

Bell, P. R. (ed.) (1959). Darwin's biological work; some aspects reconsidered; Cambridge University Press; reprinted 1964 as a paperback by Science Editions (J. Wiley); xiii + 343 pp.

Elandt-Johnson, R. C. (1971), Probability models and statistical methods in genetics; Wiley, xviii + 592 pp.

Gause, G. F. (1934), The struggle for existence; Williams & Wilkins; reprinted as paperback, 1971, by Dover; ix + 163 pp.

Hoem, J. M. (1972), Inhomogeneous semi-Markov processes, select actuarial tables, and duration-dependence in demography; pp. 251-296 in: T.N.E. Greville (ed.) Population dynamics; Proc. Symp. conducted by MRC at Univ. of Wisc., June, 1972; Academic Press, ix + 445 pp.

Kempthorne, O. (1957), An introduction to genetic statistics; Wiley, xvii + 545 pp.

Keyfitz, N. (1967), Estimating the trajectory of a population; pp. 81-113 in: Proc. 5th. Berkeley Symposium on Math. Stat. and Probability, held 1965 at Berkeley; Univ. of California Press.

Keyfitz, N. (1972), Population waves; pp. 1-38 in: T.N.E. Greville (ed.), Population dynamics; Proc. Symp. conducted by MRC at Univ. of Wisc., June, 1972; Academic Press, ix + 445 pp.

Kimura, M. & Ohta, T. (1971), Theoretical aspects of population genetics; Princeton University Press; Monographs in Population Biology 4, ix + 219 pp.

Laidler, K. J. & Bunting, P. S. (1973), The chemical kinetics of enzyme action, 2nd. ed.; Oxford, Clarendon Press; xi + 471 pp.

Lotka, A. J. (1939), On an integral equation in population analysis; Ann. Math. Stat. 10, pp. 144-161.

Mendel, G. J. (1865), Versuche über Pflanzenhybride; Verhandlungen des naturforschenden Vereins in Brünn 4; translated into English on pp. 1⁴ 144 in: A. P. Suñer, (English translation by C. M. Stern), 1955, Clas sics of Biology; New York, Philosophical Library, x + 337 pp.
Needham, A. E. (1964), The growth process in animals; Van Nostrand, xi + 522 pp.
Sheps, M. C. & Menken, J. A. (1973), Mathematical models of conceptior and birth; Univ. of Chicago Press, xxiii + 428 pp.
Usher, M. B. (1972), Developments in the Leslie matrix model; pp. 29-6 in: J. N. R. Jeffers, Mathematical models in ecology; Symp. Brit. Ecc Soc. 12; Oxford, Blackwell.
Van't Hoff, J. H. (1875), Sur les formules de structure dans l'espace; Archives néerlandaises des sciences exactes et naturelles 9, pp. 445-4

The volume ends, appropriately with Professor Williams speculation about the future role of mathematics in biology. She concludes that a new calculus must be developed for biology, one which will supplant the Leibnitz/Newton calculus whose development was motivated by the physical sciences.

NEEDS FOR THE FUTURE: RADICALLY DIFFERENT TYPES OF MATHEMATICAL MODELS

Mary B. Williams

Department of Philosophy, Ohio State University
Columbus, Ohio 43210

Since parts of future biomathematics which involve further development along directions already predominant can be extrapolated from the other papers in the volume, the purpose of this paper is to investigate less obvious, and therefore more speculative, directions. In particular, I intend to show reason to suspect that the laws of biology are of a different mathematical form than the laws of physics and that, therefore, the mathematical models of the future in biology will be radically different from the models used in physics. I will primarily concentrate on casting doubt on the ultimate usefulness of many differential equations models by examining the extent to which the mathematical assumptions underlying these models reflect biological reality. Since these mathematical assumptions seem to be seriously discordant with important parts of biological reality, this type of mathematical model should ultimately be replaced by essentially different types of mathematical models.

The mathematics used in physics was developed by physicists, and physically sophisticated mathematicians, for physics. Both the reason for the importance of differential equations in physics and the effect of this importance on our expectations is revealed in the following statement: "All scientific laws are differential equations with respect to time (at least all the laws of physics are)" (Bergman, 1957). The laws of physics _are_ differential equations with respect to time; the problem to be addressed in this paper is whether the laws of biology are.

There are two things I need to do in order to make my point. The

first is to make clear what we are trying to make a mathematics _for_
when we try to mathematize biology. In order to do this I shall first
give a concrete example which will illuminate the difference between
biologically relevant and biologically irrelevant questions, and I
will then discuss the nature of biological laws to show how the structure of biological laws can be different from the structure of physical laws even if biology is ultimately reducible to physics. The
second thing I need to do is to show some important differences between the structure of the laws of physics and the structure of the
interesting biological phenomena. In order to do this I will give
reason to believe that elapsed time is not, in general, a relevant
variable in biological laws, and reason to believe that the most
important biological 'distances' do not satisfy the metric properties
and that, therefore, the calculus is not universally applicable.

Phrasing Questions with and without Differential Equations

When we approach a subject with the expectation that differential
equations with respect to time are an appropriate tool for stating its
laws, we naturally ask questions about the rate of change of the variables over time. Any change over time can be described by such an
equation; for example, suppose that a scientist wants to use differential equations with respect to time to describe the growth of an
insect larva. He discovers that the larvae of the species in question
grow at different rates, depending on available food, temperature, and
other stochastic factors. So he forces the larvae to grow at the same
rate by making the amount of food available constant, making the temperature constant, etc. The resulting differential equation tells him
something about growth of the larvae, but it certainly does not express
a law of nature about the growth of larvae in the same way that "velocity at time t equals gt + v" expresses a law of nature about falling
bodies. The equation about larvae has ignored relevant variables. At
best it is a law which may be useful when laws describing the effects

of variable nutrition, variable temperature, etc. can be found. But if, for example, the larvae normally used cues from their food supply to hasten or retard their maturation to ensure that the adult would emerge when the food plant was in flower, this equation would be irrelevant for understanding biology. Thus the fact that a phenomenon occurs over time is not sufficient to ensure that time is a relevant variable in significant laws which elucidate that phenomenon.

Notice that in the above example there was no suggestion that the biologist had noticed that there was an interesting relationship between growth and elapsed time and then investigated it; let us look now at an example in which biologists discovered an interesting relationship and then tried to find a law which expressed it. In the middle nineteenth century many species of butterflies and moths contained mostly light-colored organisms, with a few dark-colored (black, dark brown, very dark gray) organisms in each generation. Today, in industrialized areas (though not in rural areas) these species contain mostly dark-colored organisms, with a few light-colored ones in each generation. This has happened with 80 species in industrialized parts of Britain, with over 100 species in the Pittsburgh area, and with similarly numerous species in industrialized parts of Canada, Germany, Yugoslavia, Czechoslovakia, and Poland (Ford, 1964). It seems clear that there must be a law of nature determining this ubiquitous relationship between industrialization and color change. Carefully controlled experiments in both field and laboratory have revealed that the light-colored organisms are protected from predation by visually hunting birds when they are resting on unpolluted (light-colored) tree trunks, but are subject to heavy predation when they are resting on tree trunks blackened by soot. Conversely, the dark-colored organisms are subject to heavy predation when resting on unpolluted tree trunks and are protected from predation when resting on soot-blackened tree trunks. The law might be stated as follows:

L$_B$: If (1) S is a species which relies on a resemblance to background G against visually hunting predators, (2) the appearance of the background changes from G to G', and (3) some
organisms of S are better camouflaged on G' than the typical
organism of S are, then the species S will change so that its
typical organism has the appearance which provides better
camouflage against the new background.

The crucial difference between these two problems of larvae growth
and industrial melanism is that the melanism question was suggested by
nature while the growth question was suggested by the scientific tradi-
tion, inherited from physics, of studying rates of change over time.

Is This Vitalism?

How is it possible for biological laws to have a different mathematical structure than physical laws if each biological event is also
a physical event? Consider the law L$_B$ stated above; this is a biological law because the laws which are used in proving that the consequent
is implied by the antecedent are the laws of Darwinian natural selection. (This law has not been formally proved as a theorem in an axio-
matized statement of evolutionary theory, but it seems clear that it
does follow from evolutionary theory.) It is heuristically fruitful
since it leads to biologically interesting generalizations and spe-
cializations: e.g., it could be generalized to cover non-visual as
well as visual predator deception techniques, or specialized to cover
in more detail the specific types of visual deception techniques.

Now consider the task of stating this as a physical law — that is,
of stating the conditions and the result in a form so that the result
can be derived from the conditions with the aid of physical laws. It
would be necessary to specify all of the different physical processes
by which the appearance of the organism may be determined: e.g., an
organism may be a certain color because of the amount of melanin in
its skin, because of the distribution of melanin in its skin, because

of the light-absorbing structural properties of its feathers, because it dyes its feathers to match the local ground by taking frequent dust baths, etc. Not only can one effect be achieved via many different physical pathways, but also one physical pathway may achieve different effects: e.g., an increase in the amount of melanin may cause organisms in species S' to appear darker, while the same change in amount has no effect on the appearance of organisms in species S". It may seem that all of this description in terms of the physical causes of the effect could be replaced by a simple physical description of the effect in terms of wavelengths of reflected light, but that would work only if the visual systems of all predators function in the same way, and they don't; e.g., some predators are colorblind, some see only moving objects, some are binocular, etc. Therefore it would be necessary to specify for each prey species the way in which the appearance change affects the visual system of its predator.

Thus the physical law corresponding to L_B will look like:

L_P: If species S is A_1, or A_2, or ..., or A_{674}, then species S is C. (Species S is A_1, might be: S is a species in which melanin is distributed in way D_1, whose skin structure is D_2 whose predators' visual system is ....)

There are two ways in which L_P is inferior to L_B. Firstly L_P is enormously more complicated to state, prove, and apply. Secondly L_P does not suggest fruitful generalizations; indeed the very structure of L_P suggests that further field investigation will reveal new conditions, A_{675}, A_{676}, etc., which also cause species S to be C. Since virtually every significant biological property is developed in different species by different physical pathways, the project of describing lawlike relationships among these properties in terms of the physical pathways underlying them seems likely to frequently lead to laws like L_P.

This is not vitalism. The claim is not that biological laws cannot be reduced to physical laws. The claim is that the statement 'Biston betularia is black because blackness protects it against predators in a soot-blackened environment' is deeper, more powerful, and more fruitful than the statement 'Biston betularia is black because it produces a large amount of melanin.' In general, the claim is that the structure of the deep, powerful biological laws is determined by the structure of the laws of evolution rather than by the structure of the laws of physics.

The Structure of the Laws of Evolution

The structure of the laws of Darwinian theory is specified in some detail in the axiomatization given in Williams (1970). I shall here roughly indicate enought of that structure to provide a glimpse of some aspects of evolutionary theory which bear on its mathematical structure. I will present the main points of the axiomatization very informally; because definition of the technical terms used in the later axioms would require an inordinate amount of space, they will be stated in informal translations which do not render their full meaning.

There are two sets of axioms. The first set delineates properties of the set B of reproducing organisms on which natural selection works. The primitive terms introduced in this first set are biological entity and ⊁ . ⊁ is an asymmetric, irreflexive, and non-transitive relation between biological entities and should be read 'is a parent of'. Some possible interpretations for 'biological entity' are organism, gene, chromosome, and population; in this paper I shall use only the organism interpretation.

Translation of Definition B1: b_1 is an ancestor of b_2 if and only if b_1 is a parent of b_2 or there exists a finite, non-empty set of biological entities b_3, b_4, $\cdots$, b_k such that b_1 is a parent of b_3, b_3 is a parent of b_4, $\cdots$, and b_k is a parent of b_2.

<u>Translation</u> <u>of</u> <u>Axiom</u> <u>B2</u>: For any b_1 and b_2 in B, if b_1 is an ancestor of b_2, then b_2 is not an ancestor of b_1.

The primary purpose for stating this axiom is to enable the concepts of <u>clan</u> and <u>subclan</u> to be defined. A clan is a temporally extended (over many generations) set of related organisms.

<u>Translation</u> <u>of</u> <u>Definition</u> <u>B7</u>: The clan of a set S of organisms is the set containing S and all of its descendants.

A crucial biological phenomenon occurs when a part of a clan becomes isolated from the rest of the clan (perhaps by an earthquake) and is subjected for many generations to different selective pressures until ultimately the organisms in one part of the clan are so different from their contemporaries in the other part that they would be described as different species; this is a situation of great interest for evolutionary theory, so it is clearly important to have a word for some particular kinds of parts of a clan. The first one I define is <u>subclan</u>, which, roughly speaking, is either a whole clan or a clan with one or several branches removed. Next the term <u>Darwinian</u> <u>subclan</u> is introduced; a Darwinian subclan is a subclan which is held together by cohesive forces so that it acts as a unit with respect to natural selection. Notice that in describing the term 'Darwinian subclan' I am forced to mention the theory of natural selection; this is because <u>Darwinian</u> <u>subclan</u> is a primitive term of the theory. The fourth primitive term of the theory is <u>fitness</u>; fitness is a real-valued function on the set of organisms, and is (intuitively) a measure of the quality of the relationship between the organism and its environment (which is determined by such factors as fertility, ability to get food, ability to avoid dangers, etc.).

<u>Translation</u> <u>of</u> <u>Axiom</u> <u>D2</u>: There is an upper limit to the number of organisms in any generation of a Darwinian subclan.

(This limit is important in inducing the 'severe struggle for life' which Darwin noted.)

Translation of Axiom D3: For each organism b_1 there is a positive real number which describes its fitness in its environment.

Before giving the next axiom I must introduce another term denoting a particular type of subclan. Suppose that at a particular point in time the Darwinian subclan D contains a subset of organisms with a hereditary trait that gives them a selective advantage over their contemporaries, and suppose that this trait continues to give a selective advantage for many generations. Then members of this subset will, on average, have more offspring than their contemporaries, and if we consider the sub-subclan derived from this set of organisms (call it D'), then in the offspring generation the proportion of the members of D which are in D' will be larger than it was in the parent generation. Similarly, since this trait continues to give a selective advantage, the proportion of D' will continue to increase in subsequent generations. It is by this increase in the proportion of organisms with particular traits that characteristics of populations (and species) are changed over time. It is the expansion of the fitter sub-subclan which causes descent with adaptive modification.

Translation of Axiom D4: Consider a sub-subclan, D', of D. If D' is superior in fitness to the rest of D for sufficiently many generations (where how many is sufficiently many is determined by how much superior D' is and how large D' is), then the proportion of D' in D will increase during these generations.

The final axiom asserts the existence of sufficiently hereditary fitness differences.

Translation of Axiom D5: In every generation m of a Darwinian subclan D which is not on the verge of extinction, there is a sub-subclan D' such that: D' is superior to the rest of D for long enough to ensure that D' will increase relative to D; and as long as D contains biological entities which are not in D', D' retains sufficient superiority to ensure further increases relative to D.

Is Time a Relevant Variable?

Time is a primitive term of Newtonian mechanics; in fact the axioms of Newtonian mechanics (i.e., Newton's laws) can all be expressed
as differential equations with respect to time. But time is not a
primitive term of the axiomatization sketched above. Can time be a
defined term in this system? The 'is a parent of' relation allows the
separation of organisms into different generations; can 'elapsed time'
be defined in terms of how many generations have occurred? No, because
generation time is not related invariantly to elapsed time. I see no
way of defining elapsed time within the axiomatization; there are, in
fact, both theoretical and empirical reasons for believing that an invariant relationship between elapsed time and any important biological
property would be selected against, which indicates that this undefin-
ability is not merely a property of this particular axiomatization but
is a property of the theory itself.

The theoretical reason for believing that such an invariant relationship would be selected against is that, since elapsed time is
not invariantly related to important conditions affecting organisms'
lives, any mutations which allowed the timing of a biological process
to be determined by the significant ambient conditions rather than by
elapsed time would be selected for. Thus, e.g., in the desert a mutation which caused a seed to germinate when it is wet, rather than when
a fixed amount of time has passed since its formation, will ensure that
the seed germinates under circumstances more favorable to its survival.
Similarly a gene which ensures that a certain fraction of the seeds of
an annual plant will germinate in the second spring after their formation, rather than in the first, will prevent the extinction of the line
in years of drought. We would, therefore, expect selection, rather
than physical constraints, to have determined the timing of most biologically significant processes.

Empirical evidence strongly supports this conclusion. The timing

of virtually every important chemical reaction in living matter is
determined not by 'raw' chemical-physical constraints but by enzymes;
and the presence of enzymes with those particular rate-determining
properties is the result of natural selection. Developmental processes
are typically initiated either by environmental cues or by a biological
clock which has been formed by natural selection to synchronize the
fruition of the process with the presence or absence of biologically
significant factors (pollinators, predators, rain, etc.). In addition
to these virtually ubiquitous traits, there are numerous examples of
the evolution of specific traits which allow organisms to escape from
dangerous dependence on elapsed time. Thus it seems very improbable
that there are important biological phenomena which are functions
solely of elapsed time.

One, at first sight disturbing, consequence of this lack of dependence on elapsed time is the fact that a theory without laws containing the variable 'time' cannot be used to make predictions of the
form 'Given the state of the system at time t_o, predict the state of
the system at time $t_o + t$.' This is disturbing at first sight because
our scientific background makes us expect all predictions to be of that
form, and so the claim that biological laws do not depend on elapsed
time seems to imply that biology can't make any predictions. Indeed
the fact that evolutionary theory is frequently accused of not being a
real science, on the grounds that it cannot make predictions, is (when
taken together with the common assumption that all predictions are of
that form) indirect evidence for the claim that time is not a relevant
variable in evolutionary theory. (Fortunately the assumption that all
predictions are of that form is false; a discussion of the predictions
made by evolutionary theory can be found in Williams (1973).)

Appropriateness of Differential Equations

It seems clear, then, that biological laws need not be differential equations with respect to time. Are they differential equations

at all? Is calculus an appropriate tool in biology?

First a disclaimer: Even if the laws of biology are not all differential equations, this would not imply that differential equations are useless in biological research. It would imply that differential equations must be used with great care, and that a more appropriate mathematics will ultimately supplement them. The present use of differential equations models might be compared to the use of a sledgehammer to crack pecans; someone who deeply understands both his tools and the objects he is using the tools on can produce worthwhile results with clearly inappropriate tools. A part of the job of a mathematical modeler is to produce worthwhile results with whatever tools are available; another part of his job is to produce appropriate tools.

The first problem is continuity; if the phenomena are not reasonably continuous then calculus is not an appropriate tool. That biological phenomena may not be sufficiently continuous to warrant the use of calculus is widely recognized as a problem, so I shall not here repeat the facts which indicate this. I do, however, want to reply to one argument which is frequently given as a reason for not worrying about it — the argument that physical phenomena are not really continuous either, and yet the use of calculus in physics is warranted. One might as well argue that since it works for physics, which isn't really continuous, it will work for _any_ discontinuous phenomena! If the assumption of continuity has any function in calculus, then there must be some degree of discontinuity which will prevent us from using calculus. There are, unfortunately, no rigorous criteria specifying the allowable degree of discontinuity. Discontinuity is an essential and important feature of some of the phenomena covered by biological laws: consider, for example, the difference between Mendelian (particulate) heredity and blending heredity; this important theoretical difference is a difference between a phenomenon which is essentially discontinuous and a phenomenon which is essentially continuous. Clearly the assumption of continuity must be made with great care when dealing with biology.

Continuity is only one of the assumptions that must be satisfied if the calculus is to be used. Another such assumption is that the distance involved is a metric. There are important natural 'distances' in biology; for example, the distance of one population from another in amount of relatedness. (Unfortunately 'distance' has a connotation of "how far apart" while 'relatedness' has a connotation of "how close together". This is just a verbal problem, though, not a mathematical problem.) To investigate the mathematical structure of this 'distance' let us define the degree of relatedness between a and b as follows:

$$r(a,b) = 0 \quad \text{if a and b are full siblings}$$
$$= 1 \quad \text{if a and b are first cousins}$$
$$= 2 \quad \text{if a and b are second cousins}$$
$$\text{etc.}$$

Then r does not satisfy the triangle inequality; that is, it is not true that for all a, b, and c, $r(a,b) + r(b,c) \geq r(a,c)$. It is easy to give a counterexample to the triangle inequality for relatedness: Suppose John is a first cousin of Mary and Mary is a first cousin of Betty, then $r(\text{John, Mary}) + r(\text{Mary, Betty}) = 1 + 1 = 2$; for the triangle inequality to be satisfied, John would have to be at least a second cousin of Betty. John, however, need not have any ancestors in common with Betty, since it may be that John was related to Mary through Mary's mother and Mary was related to Betty through Mary's father. Now the crucial factor allowing this to not be a metric is that every person has two parents; perhaps if we restrict the concept to relatedness between two species then relatedness will be a metric, since a species does not have two parent species (except, occasionally, plant species). Unfortunately relatedness between two species is a function of two things: genetic relatedness from common ancestors and environmental relatedness from similar selection pressures. Thus, for example, turtles are considered to be more closely related to crocodiles than to birds, although by the sole criterion of common ancestry turtles are more closely related to birds. Those two independent factors entering into relatedness between species will generate the same type of counterexample to the triangle inequality as was given above for relatedness within species. So this most important natural biological distance is not a metric.

These examples cannot provide definitive evidence that differential equations are not appropriate tools for expressing important biological laws; they are merely indicative. But they indicate to me that we should look elsewhere for a mathematics for biology.

Hints for the Future

But where can we look? There is one small indication that can be described of what some of the important mathematics for biology is. In the two major integrating theories of biology (Darwinian theory and Mendelian theory) the 'is a parent of' relation is important. This relation determines a strict partial ordering on the set of biological entities; therefore it determines a graph, and we might hope that graph theory will provide a mathematical home for some biological laws. But graph theory will not provide a ready-made mathematical theory for biology; the Darwinian graphs are non-planar and non-finite, while virtually all the graph theory that has been developed is about finite and planar graphs. Still, graph theory provides an indication of a type of mathematics which is relevant.

Another indication of a possible direction is given by the fact that the discontinuous units important in biology make the vector notation natural in the description of many biological phenomena: the resource vector for a species (the set of species on which the given species feeds); the feeding tactics vector for a species (the probabilities of the given species eating its different resource species); the age-specific survival vector for a species (the probabilities of survival at different ages); etc. The Leslie matrix, which is used for calculating the age structure in the population in the next generation from the age structure, fertilities, and survival probabilities of the present generation, is already used by wildlife biologists, and we may expect to see a good deal more use of linear algebra. It seems likely that, in some biologically important situations, operations which do not satisfy the group properties will be used; the work already done by algebraists on such operations will provide some groundwork for the discovery of the properties of the biologically important operations, but detailed theories of these structures will have to be developed as the biologically significant properties of the relationships among the vectors are discovered.

Awareness of the vast extent of variability in the biological world suggests that stochastic models of biological phenomena will play an important role in future biomathematics. Much work in this area has of course already been done.

Conclusion

I have indicated in this paper some reason to believe that an important source of difficulty for us in developing mathematical models for biology has been that we have been trying to use the calculus, a mathematical system developed for physics and in many cases seriously inappropriate for biology. The fact that natural selection will free biological systems from a disadvantageously strong dependence on the physical properties of their building blocks makes it likely that in general the mathematical tools developed for physics will not be the best tools for biology. If the types of mathematical structures available in the contemporary mathematical toolbox are not appropriate for the biological problems that we need to solve, then it is our job to develop mathematical tools that are appropriate.

Literature Cited

Bergman, Gustav; _Philosophy of Science_; University of Wisconsin Press, Madison, WI; 1957.

Ford, E. B.; _Ecological Genetics_; Methuen (Wiley); New York; 1964.

Williams, Mary B.; 'Deducing the Consequences of Evolution: A Mathematical Model', _J. Theoret. Biol_. 29 (1970) 343-385.

Williams, Mary B.; 'Falsifiable Predictions of Evolutionary Theory', _Phil. of Science_, vol 40, no. 4, December 1973, 518-537.

Editors: K. Krickeberg;
S. Levin; R. C. Lewontin;
J. Neyman; M. Schreiber

Biomathematics

Vol. 1:

Mathematical Topics in Population Genetics
Edited by K. Kojima
55 figures. IX, 400 pages. 1970
ISBN 3-540-05054-X

This book is unique in bringing together in one volume many,
if not most, of the mathematical theories of population
genetics presented in the past which are still valid and some
of the current mathematical investigations.

Vol. 2:

E. Batschelet
Introduction to Mathematics for Life Scientists
200 figures. XIV, 495 pages. 1971
ISBN 3-540-05522-3

This book introduces the student of biology and medicine to
such topics as sets, real and complex numbers, elementary
functions, differential and integral calculus, differential equations, probability, matrices and vectors.

M. Iosifescu; P. Tautu
Stochastic Processes and Applications in Biology and Medicine

Vol. 3:

Part 1: Theory
331 pages. 1973
ISBN 3-540-06270-X

Vol. 4:

Part 2: Models
337 pages. 1973
ISBN 3-540-06271-8

Distribution Rights for the Socialist Countries: Romlibri,
Bucharest
This two-volume treatise is intended as an introduction for
mathematicians and biologists with a mathematical background
to the study of stochastic processes and their applications in
medicine and biology. It is both a textbook and a survey of the
most recent developments in this field.

Vol. 5:

A. Jacquard
The Genetic Structure of Populations
Translated by B. Charlesworth; D. Charlesworth
92 figures. Approx. 580 pages. 1974
ISBN 3-540-06329-3

Population genetics involves the application of genetic information
to the problems of evolution. Since genetics models based on
probability theory are not too remote from reality, the results
of such modeling are relatively reliable and can make important
contributions to research. This textbook was first published
in French; the English edition has been revised with respect
to its scientific content and instructional method.

Springer-Verlag
Berlin
Heidelberg
New York